TROPICAL SUSTAINABLE ARCHITE(

Social and Environmental Dimensions

TROPICAL SUSTAINABLE ARCHITECTURE

Social and Environmental Dimensions

Joo-Hwa Bay and Boon Lay Ong

AMSTERDAM • BOSTON • HEIDELBERG • LONDON • NEW YORK • OXFORD
PARIS • SAN DIEGO • SAN FRANCISCO • SINGAPORE • SYDNEY • TOKYO

Architectural Press is an imprint of Elsevier

Architectural
Press

Architectural Press is an imprint of Elsevier Ltd
Linacre House, Jordan Hill, Oxford OX2 8DP
30 Corporate Road, Burlington, MA 01803

First edition 2006

Copyright © 2006, Joo Hwa Bay and Boon Lay Ong. Published by Elsevier Ltd. All rights reserved

No part of this publication may be reproduced, stored in a retrieval system or transmitted in any form or by any means electronic, mechanical, photocopying, recording or otherwise without the prior written permission of the publisher

Permission may be sought directly from Elsevier's Science & Technology Rights Department in Oxford, UK: phone (+44) (0) 1865 843830; fax (+44) (0) 1865 853333; email: permissions@elsevier.com. Alternatively you can submit your request online by visiting the Elsevier web site at http://elsevier.com/locate/permissions, and selecting *Obtaining permission to use Elsevier material*

Notice
No responsibility is assumed by the publisher for any injury and/or damage to persons or property as a matter of products liability, negligence or otherwise, or from any use or operation of any methods, products, instructions or ideas contained in the material herein. Because of rapid advances in the medical sciences, in particular, independent verification of diagnoses and drug dosages should be made

British Library Cataloguing in Publication Data
A catalogue record for this book is available from the British Library

Library of Congress Control Number: 2006927680

ISBN 13: 978-0-75-066797-5
ISBN 10: 0-75-066797-4

For information on all Architectural Press publications
visit our website at www.books.elsevier.com

Typeset by Cepha Imaging Pvt Ltd, Bangalore, India
Printed and bound in Italy

06 07 08 09 10   10 9 8 7 6 5 4 3 2 1

**Working together to grow
libraries in developing countries**

www.elsevier.com | www.bookaid.org | www.sabre.org

ELSEVIER   BOOK AID
              International   Sabre Foundation

CONTENTS

FOREWORD

Often, a group of individuals may come together from different disciplines in a conference to share their expertise and knowledge around a particular topic or idea. Less often, a conference proceedings or publication follows. In rare instances, the published proceedings contain sufficient relevance and substance to be a touchstone of significant development of those ideas well into the future. This publication is such a volume.

The authors and contributors to this volume raise relevant questions and offer significant and promising answers of how best to build in the tropical climates. This publication frames the terms for a continuing productive investigation of these ideas. This volume makes a very significant contribution in defining unique qualities and contributions of tropical sustainable architecture.

The terms of sustainable development and design have been variously defined and are now well known as part of the international dialogue to protect earth's resources, defined as equal portions of social equity, economic opportunity and environmental responsibility. These goals – avidly debated at the 1992 United Nations Earth Summit held in Rio de Janeiro – are essentially a call to "future fairness." Like all great architecture – sustainable architecture is that which is enduring, empowering, and inspiring, appropriate to particular climates, resources and cultures.

The theme of this volume is to define and develop what is most unique and potent in a tropical sustainable architecture. The challenge is set forth in the Preface in a set of questions by editors Joo-Hwa Bay and Boon-Lay Ong. Their questions demand an in-depth look at what it means to design and build in tropical areas in ways that are sustainable.

The contributors to this volume offer significant experience and knowledge about designing buildings in tropical climates. The chapters contain a myriad of approaches to designing buildings and urban complexes most appropriate and unique to tropical cities, regions and populations.

Some of the proposed designs and ideas are known. Others are new. Some are founded on practice, carefully documented and observed. Others are revealed by new means of analysis, including computer simulation and visualization. Each of the chapters in this volume offers new insight and knowledge useful for practice. The sum of the chapters makes the case that defining what is unique about tropical sustainable architecture is a productive focus for present and continuing research and practice.

In "The Selective Environment," Dean Hawkes describes an approach to design that is, like nature, *selectively adaptive* to specific local climate and environmental influences. Among the characteristics of Selective Environments is the notion of "free-running" buildings, a term to describe buildings where climate conditions are varied during the day and during the seasons. This variation he notes is more generally preferred over the more standardized and uniformly air-conditioned interiors. He advocates and gives examples of buildings with "buffers" which experience space, light, wind and time variations between the exposed outdoor climate and the enclosed and conditioned interiors.

In "Green Design in the Hot Humid Tropical Zone", Ken Yeang clearly summarizes what is most difficult to describe, a way of thinking about designing in the tropics, which he terms *ecodesign,* "designing the built environment as a system within the natural environment." Yeang makes it simple: "We start by looking at nature." He argues that this way of "holistic thinking" extends to the building surround, its context in physical, social and economic terms, recognizing the client and business of a building as essential partners.

Most of the ensuing chapters investigate particular topics and elements of architecture of the tropics. Joo-Hwa Bay and co-authors offer a careful study of verandas in high-rise buildings, which indicates the social amenity and climatic variability and comfort offered by well-designed veranda configurations. Their analysis demonstrates that well-designed spaces in and around buildings provide benefit by even the slightest variation of light, shade, temperature, breeze and greenery. Their design guidelines exemplify the translation of complex research into terms easily understood and applied to practice.

Boon-Lay Ong and Chi-Nguyen Cam add to the microclimatic level of discussion in their investigation of high-rise buildings. Their research indicates that improvements for comfort, such as natural lighting and ventilation, can be achieved by thoughtful design – they too use the term *holistic design* – without adding to cost.

This highlights the significant role of the designer, who is able to make decisions about building plans and placement that determine the specific microclimatic benefits (or liabilities) of a building. This point alone supports their premise that sustainable design depends on the decisions made by the building designer and, in turn, the building sponsors and clients who hire the designer. It is to this group that various sustainable building evaluation systems, such as Building Environmental Assessment Methods (BEAMs) have most to offer as design guidance and incentive.

Hsien-Te Lin, in a chapter about green building policy and evaluation, presents important design guidelines for building in the tropics, paying particular attention to building materials and components of design and construction. His conclusions describe the importance of solar control and ventilation in tropical regions, in contrast to insulation strategies so critical in other climatic zones. He notes that the vernacular elements of tropical architecture offer some approaches, "rectangular layout, short indoor depth, two side opening, veranda, and deep shading roofs." He extends these to additional factors, including ecological biodiversity, energy conservation, waste reduction, and health that now define the broad reach of sustainable design. He offers health and environmental impact data to show how building design and construction help influence if not determine these qualities. His interest is urban- and region-wide, so often seen as being beyond the designer's reach. Yet his examples show design solutions that are easily realizable by design: permeable pavement, constructed wetlands in housing estates, shading, and ventilative openings in building.

Q.M. Mahtab-uz-Zaman and co-authors provide a study of building and planning regulations. They illustrate sustainable urban strategies realizable by small adaptations of regulations and by decisions well within the capability of architects and builders. Their illustrative case is exemplary. It demonstrates the significance of a single planning factor, the floor area ratio (FAR). Varying the height and placement of buildings within city blocks can vastly improve health and environmental conditions as well as social amenities. Their illustrations remind one of "weeding one's garden" to allow light and air to regenerate life, in this case, for neighborhoods and urban residents.

Edward Ng and co-authors offer a study with a different set of terms and context, but leading to nearly identical conclusions: in tropical humid climates, varying building heights in dense urban environments provides the greatest range of environmental benefits of shade, reflected light, and ventilation. They also document that in a dense high-rise urban complex, the units and spaces closest to the ground receive least benefit of climatic comfort and variation. This prompts their design investigation of

urban morphology to provide more light and air to the street level –
resulting in images that are again reminiscent of "weeding one's
garden." These studies are particularly interesting because they
demonstrate distinctive urban forms uniquely suited to tropical
humid climates. They provide a splendid comparison to studies by
Ralph Knowles, published elsewhere, of urban morphologies and
form in northern climates generated by solar orientation criteria.

Nyuk-Hien Wong and Yu Chen continue this rich vein of research
in a study of urban heat island effects. Their findings also support
the significant role that planning and building design decisions have
on urban climate and the benefit served by good design, including
the geometry of building shape and orientation, and green spaces
and surfaces. Their analysis is fine-grained: a discussion of "cold
materials" describes the significant impact of building materials that
are selected for building exteriors. Even an apparently small decision
on the color, texture and material of the building surfaces that the
designer selects makes a difference.

Elias Salleh provides additional documentation and design guidelines based on microclimate within urban complexes, underlining
further the value of solar orientation and ventilation studies in
design. Particularly useful are his recommendations of street geometry, indicating the benefit of some ratios and orientations of streets
(*urban canyons*), where spacing and orientation can make or break
a sustainable outcome. A gardener will talk of the critical role of light
and air on the health of plants, particular to their spacing and position in the garden. Here Salleh gives evidence of the critical role of
measurement in designing for urban health, comfort and well-being.

In separate chapters, Shoichi Ota and Jose Roberto Garcia Chavez
each describe prototype house designs that experiment with building design and materials. Ota's proposal for the old quarter in Hanoi
is a fully realized prototype, using an approach he calls *porous structure*, by which houses, courtyards and even individual rooms are
offered light and air in relatively dense arrangements. The design,
built as a model house, is resolved down to the level of sun screen-
ing, interior light and spaces. The restraint and elegance of the
design speaks of the inspirational promise of architecture.

Garcia Chavez offers a remarkable alternative to industrialized
housing production with a design that recycles glass bottles and utilizes self-help and mutual aid construction techniques for low-cost
housing. The model of this design has been built as part of an affordable and sustainable community building, including the women and
children who built the model house, and integrated with family and
community gardening, aquaculture, and recycling. Here is an example of holistic/sustainable design realized as holistic/sustainable
community development.

Much to the credit of the editors, the presentation of ideas is balanced by a healthy overview of evaluation and critique. An essay by Alexander Tzonis is presented at the beginning of this volume and places the discussion of sustainable design amongst a complex set of trends in the evolving history of ideas in architecture. In this perspective, the emphasis given here to climate design and environmental measures is part of a continuing mix of ideas, influencing and influenced by architectural theory and practice in rich and unpredictable ways.

The concluding chapter by Anoma Pieris – best read as intended at the end and after reading all others – is cautionary and thus salutary. It is cautionary because the author argues that concepts of design approaches based on "invocation of climate, local geography, location materials and construction [has] generated a mirage of an environmentally sustainable practice." The author thus urges a continuing self-awareness and critique to be true to the word (promise) of sustainable design. She ends her chapter, and thus of this volume, with a provocative statement, that will serve well if it incites more thought, discussion and action.

And thus with this entire volume, it deserves to be read, taken up as the touchstone and provocation of more research and practice to pursue the vital tasks of designing and building appropriate to the tropical climate and cultures. The entire world can be inspired by this auspicious, rich and reasoned beginning.

Donald Watson

DONALD WATSON is an architect and educator, former Dean and Professor Emeritus of Rensselaer Polytechnic Institute. He is the author of numerous books, including *Building Climatic Design* (1984) and *Energy Design Handbook* (1999), and is editor of *Time-Saver Standards of Urban Design* (2003). Among his recognitions and awards are the PLEA Lifetime Achievement Award (1990), AIA Education Honors Award (1997), ACSA Distinguished Professor Award (2003) and ARCC James Haecker Leadership Award in Architectural Research (2005).

PREFACE

The tropical belt – where large areas of South East Asia, India, Africa, and parts of both North and South America are located – forms the biggest landmass in the world and has one of the highest number of rapidly developing cities. Perhaps coincidentally, architecture in the tropical regions share common problems, of which perhaps the most easily identified, is the tropical conditions of climate and natural environment. The context for architecture in these regions is fraught with conflicts between tradition and modernization, massive influx of the rural poor into urban areas, poorly managed rapid urban development, the cultural and social strain of globalization and other issues as yet undefined. There are many questions concerning tropical sustainable architecture that involve both social and environmental dimensions:

- What design strategies are suited for tropical high-density city living of many rapidly growing urban hubs that take local environment and social-cultural needs into consideration?
- Do imported technologies, skills and knowledge engage with the diverse local cultural traditions and lifestyles of the tropical regions? Do they optimize natural environment and the established and evolving cultural habits for maximum energy savings?
- What comfort indices and environmental standards are suitably developed for the planning and design for tropical conditions and lifestyles?
- Are there local and traditional methods and resources for planning and building, linked to the established ways of living that can be adapted for contemporary sustainable developments?
- Is there a marked separation of research in cultural studies and environmental parametric studies? Are there more holistic solutions for sustaining culture and environment?

The recent conference of the International Network for Tropical Architecture, iNTA 2004 held in Singapore, with submissions from over 24 countries, succeeded in laying the foundations for future discussion on tropical architecture as well as presenting a broad canvas of latest research and thinking in this area. We take this opportunity here to acknowledge the kind help of the fellow members of our organising committee, namely Johannes Widodo, Nyuk-Hien Wong, Swee-Ling Tse and Lai-Choo Lee, and Malone who worked with us in making this conference possible, and publishing the conference proceedings, entitled "1st International Tropical Architecture Conference, Architecture and Urban Design in the Tropical Regions: Sustainability and Society".

This book, *Tropical Sustainable Architecture: Social and Environmental Dimensions*, different from the proceedings, presents an array of voices that address these issues directly and in passing. Several speakers from the conference were invited to write fresh essays for this book, relating to the theme. The views may come from specialized fields that in themselves do not speak of the larger context but when placed together begin to provide a faceted vision of the future of architecture in the tropics. But there are also essays that directly address the problems facing tropical architecture. The theoretical frameworks, research and practice solutions discussed in this book are pertinent for the improvement in the environmental and social sustainability in the tropical regions.

J.H. Bay and B.L. Ong

ABOUT THE AUTHORS

Joo-Hwa Bay has been practising in the tropical region since 1986. He received his Ph.D. from Technische Universiteit Delft (TUDelft) on design thinking in tropical architecture. He has been the director of a large practice and a Council Member of the Singapore Institute of Architects, and has won several design awards. He currently teaches and researches at the Department of Architecture, National University of Singapore. His published works include *Contemporary Singapore Architecture*, *Cognitive Biases in Design: The Case of Tropical Architecture*, and a chapter on 'Three tropical paradigms' in *Tropical Architecture: Critical Regionalism in the Age of Globalization*. He has been invited to speak in many international conferences and seminars. He chairs the International Network for Tropical Architecture (iNTA).

His co-authors of Chapter 5 in this book **Na Wang**, **Ping Kong**, and **Qian Liang** were his research assistants at the National University of Singapore. They completed their Masters dissertations on inter-related topics supervised by Bay. Wang and Liang are currently practicing architecture in USA and Singapore, respectively, and Kong is pursuing a Ph.D. at TUDelft.

José Roberto García Chávez is a lecturer in Architecture at the Metropolitan Autonomous University of Mexico City and has been a visiting lecturer in several universities worldwide. He realized his architecture degree in Mexico and Ph.D. in the United Kingdom. He is involved in teaching, research and professional practice on Bioclimatic Architecture, Renewable Energies and Sustainable Development applied in buildings as well as in rural and urban ecological communities. He has written more than 90 papers presented in international conferences and journals and also authored or co-authored several publications on his field.

Dean Hawkes is a fellow at Darwin College, Cambridge. He currently holds visiting professorships at the University of Huddersfield and the Glasgow School of Art, UK and is honorary visiting professor at the Chinese University of Hong Kong. In 2000 he was honoured with the PLEA award in recognition of his contribution to the teaching, research and practice of passive, low-energy architecture. He currently holds a Leverhulme Emeritus Research Fellowship to study 'The Environmental Function of Architecture'. His most recent books are *The Environmental Tradition, The Selective Environment*, with Jane McDonald and Koen Steemers, and *Engineering and Environment*, with Wayne Forster. His works in practice are also well published and exhibited.

Hsien-Te Lin is the chairman of the Green Building Committee of Taiwan and a professor at the Department of Architecture, National Cheng Kung University, Taiwan. He received his Ph.D. from Tokyo University, on Dynamic Building Energy Simulation and Green Building Design. He received the best annual research paper award by the Air Conditioning and Sanitary Society of Japan in 1987. He created a simplified index – ENVLOAD for building energy conservation which has become the National Building Energy Conservation Code of Taiwan since 1995. He also developed a Green Building Evaluation System, EEWH, regarded as the official sustainable standard and building policy in Taiwan since 1999.

Edward Ng is a professor at the Department of Architecture, The Chinese University of Hong Kong. His co-authors of Chapter 9 are Tak-Yan Chan, Vicky Cheng, Nyuk Hien Wong, and Meiqi Han. **Tak-Yan Chan** and **Vicky Cheng** are colleagues of Ng, while **Nyuk-Hien Wong**, an associate professor, and **Meiqi Han** are colleagues at the Building Department, National University of Singapore. They are all collaborators in their research project on Designing High Density Cities Parametric Studies of Urban Morphologies and their Implied Environmental Performance.

Boon-Lay Ong is a senior lecturer at the Department of Architecture, National University of Singapore. He received his Ph.D. from Cambridge University, UK on the use of plants in the making of place in architecture. His contributions in this field include FOLIAGE – a computer program to simulate the thermal effect of plants on building, the Green Plot Ratio – a metric for greenery in architecture and urban design, and the concept of Nurtured Landscapes – the use of designed landscapes as an ecological element in architecture and urban ecosystems. His ideas on integrating plants and landscape into architecture have bagged two design awards and the Green Plot Ratio is currently being considered for implementation in Singapore. His work has been published in international textbooks.

His co-author of Chapter 6 is **Chi-Nguyen Cam**. Cam has successfully completed his Ph.D. thesis at the Department of Architecture, The National University of Singapore, on the validity of building environmental assessment methods (BEAMs) for sustainable performance of high-rise housing in Singapore under the supervision of Boon-Lay Ong. He is currently practicing architecture in Singapore.

Shoichi Ota is teaching and researching at the Institute of Industrial Science, University of Tokyo, Japan. His Ph.D. dealt with the modern history of architecture and city planning in Vietnam. He once served as assistant lecturer at the Hanoi University of Civil engineering. His research interests are in architectural history, heritage studies, and urban studies in Asia.

Anoma Pieris is a lecturer in Architecture at the University of Melbourne. She has architecture degrees from the University of Moratuwa (Sri Lanka), the Massachusetts Institute of Technology and obtained her Ph.D. from University of California, Berkeley. Her research interests include the history of Modern South and Southeast Asia and the construction of specific architectural discourses within the region. She is the co-author of *New Directions in Tropical Asian Architecture* and the author of *JCY: the Architecture of Jones Coulter Young*.

Elias Salleh is a Malaysian professional architect with an extensive working experience of over 30 years in local architectural education. He has served in various academic and management positions at Universiti Teknologi Malaysia (UTM) since 1973 before serving as a deputy vice-chancellor of Universiti Utara Malaysia in 1999 until his official retirement in 2003. He is currently a professor of Architecture at the Universiti Putra Malaysia (UPM), specialising in environmental science and sustainable architectural studies. He obtained his Masters degree in Building Science from the University of Sydney, Australia, and subsequently his Ph.D. (Environment and Energy) from the Architectural Association Graduate School of Architecture in London.

Alexander Tzonis holds the chair of Architectural Theory and Design Methods at the University of Technology of Delft and director of Design Knowledge Systems (DKS). He was a distinguished visiting professor to the National University of Singapore. Among his many books are *Towards a non-oppressive Environment,*

The Shape of Community with Serge Chermayeff, with Liane Lefaivre he introduced the term critical regionalism into architecture, and wrote *Architecture in Europe since 1968, Architecture in North America since 1960*, and *Critical Regionalism*. With Liane Lefaivre and Bruno Stagno, he edited *Tropical Architecture: Critical Regionalism in the Age of Globalization*. He has also written books on the works of Santiago Calatrava and Le Corbusier.

Nyuk-Hien Wong is an associate professor and the program director of M.Sc. (Building Science) at the Department of Building, National University of Singapore. He has done extensive research in the area of Urban Heat Island (UHI) and explored the use of urban greenery to mitigate the UHI effects. He has published more than 100 international referred conference and journal papers and has published extensively in journals including Building and Environment, Energy and Buildings. His co-author of Chapter 10 in this book is **Yu Chen**. Who has successfully completed his Ph.D. thesis at the Department of Building, National University of Singapore, on the heat island effect in Singapore under the supervision of Wong. He is currently practicing architecture in Singapore.

Ken Yeang is an architect, specializing in the design and planning of 'green' or ecologically-responsive large buildings and sites; whose firm Hamzah & Yeang, has offices in London, Kuala Lumpur (Malaysia) and China. Yeang's Cambridge Ph.D. dissertation on ecological design and planning has been the agenda for his firm's design and R&D design work. The firm has received a number of international awards including the Aga Khan Award for Architecture, the Royal Australian Institute of Architects International Award. Yeang has written many books and his latest include *The Green Skyscraper: The Basis for Designing Sustainable Intensive Buildings*. He has lectured on the topic of ecological design theory and practice to audiences in over 30 countries.

Q.M. Mahtab-uz-Zaman is an associate professor of Architecture at the Department of Architecture, BRAC University. His co-authors of Chapter 8 are Jalal Ahmad, Fuad H. Mallick and A.Q.M. Abdullah. **Fuad H. Mallick** is a professor and **A.Q.M. Abdullah** is a lecturer at the same university, and both are research colleagues of Zaman. **Jalal Ahmad** is a practicing architect and a member of the Institute of Architects Bangladesh. They all are collaborators on a research project on Habitable Urban Space-Built Ratio: A Case Study of Building and Planning Regulation in Dhaka City.

1 SOCIAL AND ENVIRONMENTAL DIMENSIONS IN TROPICAL SUSTAINABLE ARCHITECTURE: INTRODUCTORY COMMENTS

Joo-Hwa Bay and Boon-Lay Ong

Department of Architecture, National University of Singapore

Keywords

Architecture, urban design, ethics and poetics, social, environment, design guidelines, high-rise high-density, tropical, sustainable, ecological.

Why tropical architecture? Critics argue that the term ought not to exist, and that it is perhaps a misnomer or a faux pas. Can any architecture built and inhabited in the tropics not be tropical? Some point out that the phrase, if not the concept, originated during colonial times and is a vestigial legacy of European sovereignty in South East Asia. Not something to be mentioned in polite company. Then there is the problem of political boundaries. Many of us practicing in South East Asia forget that tropical architecture applies also to parts of Australia, Africa and the Americas. On the other hand, the influence of traditional tropical architecture may be seen in Asia as far north as India, China and Japan – countries which are largely not tropical in climate.

This book, resulting from the first conference held by the International Network for Tropical Architecture (iNTA) in 2004, does not provide a simple answer to all these questions. It was not set out to do so. The conference itself had no lack of interested participants – garnering over 150 applications from 24 nations. iNTA[1] was itself constituted during this conference and has gone on to be staged in 2006 in Indonesia and thereafter, if all goes well, in Australia also.

The significance of tropical architecture lies beyond its climatic and regional concerns. Inasmuch as it confronts the spread of a homogenous globalism and argues for a locally and environmentally sensitive approach, it also signals the issues and contentions for a sustainable future. Indeed, the essays presented in this book are distinctive not just for their relevance to the tropical nations today but in their engagement with issues that ought to be of concern to architects everywhere. To this end, we have carefully selected presenters from the conference and asked them to write essays especially for this book[2].

1.1 TROPICAL ARCHITECTURE AND MODERNISM

One of the impetuses for developing research around tropical architecture lies in the historical origins of modernism in Europe and the US. From its early days, modern masters like Le Corbusier and Oscar Niemeyer understood that Modernism in architecture ought not to be transplanted globally without some recognition of its changed context. Alvar Aalto, in championing modernism in the Scandinavian countries, also emphasized the importance of understanding the region, climate and social context. In the US, Frank Lloyd Wright devised the term Usonian architecture to emphasize the grounding of his architecture on locality. The Arts and Crafts movement in the UK too sought their grounding in local traditions and to preserve their culture.

Unfortunately, much of the work that passes for architecture in the tropics today are unadulterated transplants from temperate countries, particularly the US – justified in the name of International Style. The inappropriateness of such transplants was argued by many schools of architecture, armed with the writings of environmentalists like Victor Olgyay (1952) with Aladar Olgyay (1963) and Maxwell Fry (1956) and Jane Drew (1964), and the designs of architects like Paul Rudolf, Richard Neutra, and several local masters, many of whom are not known outside the shores of their own countries (Lefaivre and Tzonis, 2001). The ease with which the International Style can be transplanted and the ubiquity and low-cost of energy as a solution to any environmental woe meant that by and large, the call for a more appropriate architecture went unheeded in the tropics – as it was elsewhere.

Countries in the tropical belt have seen unprecedented growth in the last 50 years and are poised to escalate in terms of economic, technological and material development. Not only are the issues facing countries in the tropical belt relevant to other countries, it is also likely that countries in the tropical belt will be among the leaders in terms of economic and urban development in the world in the foreseeable future. The fact that the impending escalated development in the tropics is unprecedented poses new problems and challenges to architects and planners all over the world and requires fresh ideas from our very best thinkers.

1.2   CLIMATE AS DESIGN GENERATOR

The point of departure for most tropical architecture is climate. At the simplest level, modern tropical architecture has been simply an adaptation of modern trends in design and construction to climate, taking into consideration some changes in the lifestyle that the tropical climate affords. Often, there has been an exploration of open and semi-open spaces, verandas and balconies, and open plans.

But the city and the modern lifestyle it encapsulates do not encourage such natural living. Strong breezes, so welcome in the hot humid conditions, mean that paper had to be kept in place or it might be blown away. Iron rusts, materials deteriorate and fungi grow faster in the tropics than in temperate countries. Working indoors, as most city dwellers do, is not comfortable and productivity is low. Most buildings in the tropical city adopt air-conditioning as the panacea to all these. The further advantage of air-conditioning is that, if costs and environmental degradation are not of concern, the architecture can be entirely stylistic and the immediate environmental and climatic conditions ignored. For these two reasons – that it can resolve environmental comfort problems by simply guzzling more energy resources, and it is stylistically open-ended – air-conditioned buildings are both popular with the public and a bane to the responsible architect.

The challenge to define a modern idiom for tropical architecture is not just a climatic issue but also one that is related to the problem of adapting to the modern lifestyle, of the transformation of local cultures to the modern city. While it is possible to retain enough of the vernacular lifestyle for residential designs to be naturally ventilated, other building typologies like offices and shopping centres have not been so lucky.

1.3   ETHICS, POETICS, SUSTAINABILITY AND CONTEMPORARY ARCHITECTURE

Pérez-Gómez (2005), in "Ethics and poetics in architectural education", asserts that the architect has a responsibility to make poetic statements of social and cultural conditions with their works. Science has its limits, and aesthetics is not just the icing on the cake or an afterthought but an intrinsic responsibility of the architect.

Unfortunately, the ease with which architects can use the umbrella of the International Style to indulge in form-making can lead to an "emotivism" (Bess, 1996) in architecture that blithely ignores all aspects of context, climate, environment and even human needs. Aesthetics in architecture becomes simply self-expression and self-referential. To give it its due, emotivism has given us some very interesting and awe-inspiring architecture – e.g. the Guggenheim Museum in Bilbao by Frank Gehry or the Sydney Opera House by Jørn Utzon. Monumentalism and iconism have their places too.

Architects have to contribute to mankind with their work. The question is how? Whatever the personal stands of individual contributors, this book supports Communitarianism. Susan Hagan (2001) proposes a moral contract between architecture and the environment. Since the 1970s, many architects, e.g. Tay Kheng Soon and Ken Yeang in Malaysia and Singapore[3], have criticized the limitations of architects who restrict themselves to the discussion of 'linguistics' and 'styling' without engaging the volatile environment and the city.

A common perception is that sustainable environments are not really "visible," and "trendy architecture" with seemingly sustainable features can fail badly and not work ecologically. Lucius Burckhardt (1992) comments that ecological architecture, or more precisely, the ecological house, an issue from the 1970s, has turned out to be a trap. They appeared well in magazines, won awards, and were examples to model after, but failed miserably in terms of measurable environmental performance. He suggests that one cannot really *see* an ecological building, but one can either build the image of an ecological house or one can calculate how to save energy and how to clean up the environment. The problem with the second option is that nobody will take photographs of it for architectural publication. Perhaps this problem can be avoided if we can bring the two dimensions together.

William McDonough (1996) hits out at irresponsible architectural practices that seek only short-term benefits, and proposes reforms with a "Declaration of Interdependence" (similar to issues raised in many fields since the first Earth Day in 1970). He claims that architects have a special role to play, where design becomes the foremost statement of human intention. He postulates that the new role of architects is one of leadership in developing new definitions and measures of prosperity, productivity, and quality of life. In the "Hannover Principles," he also proposes criteria for assessing whether a design solution is safe and just, and operate from the current solar income.

In contemporary architecture, there is a trend towards the commercialization of the image that titillates, aggravated by the internet and the flat screen. Juhani Pallasmaa (1996), in *The Eyes of the Skin*, criticized the hegemony of vision in architecture, and proposed more enduring and rich ways to sense and experience the environment and the place. Pallasmaa (1993) had also suggested that architecture will pick up on early Functionalism with a social mission, with better understanding and sophistication, shifting from the "metaphorical" towards an "ecological-functionalism." Tzonis and Lefaivre (1990), in "Critical Regionalism", traced Mumford's position that the modern architect could and should engage a place and its community critically, using innovation in technology in a progressive way, thus ensuring continuity as well as change and growth into the future. Donald Watson (1991, 1995) on rethinking good architecture, suggests that architecture that embraces sustainability issues of a context is akin to Le Corbusier's precepts for an ideal architecture.

1.4 BUILDING SCIENCE AND THE ARCHITECT

Building science is, in many ways, an attempt to reduce environmental issues and their spatial dimensions into mathematical formulas and numbers. Two threads of research can be detected here. The first deals with physiological studies and is epitomised by Ole Fanger's concept of thermal comfort. Such studies are the bedrock upon which environmentally responsive architecture can be devised. They can, and often are, legislated as design standards that provide a much-needed objective framework for architects. While it is, on the one hand, highly scientific, physiological standards of comfort do not translate readily into design (Ong, 1994, 1997; Ong and Hawkes, 1997). They are best utilized as checklists after the design has been put at least to paper.

As building science becomes more sophisticated and complicated, the danger is that the architect will find it increasingly harder to incorporate its findings into his work. While building science is reductionist and precise, in contrast, the architect's design mind tends to deal with pre-parametric heuristics, qualitative thinking with transformations from precedents, allowing him to make quick and efficient design decisions that synthesize complex criteria besides including environmental issues (Bay, 2001b). The difficulty in architectural design is that the architect has to address many diverse issues simultaneously during development of his design. The issues may be isolated, different and even conflicting but the product is a singular building that responds to all these issues at the same time. The only way for an architect to function adequately is to internalize these issues and intuitively resolve them through creative design. Architects tend to avoid applying building science directly in their design process, preferring where feasible to work with a building scientist as a consultant. The typical architect prefers to work with broad principles and strategies (Hyde, 2000). A survey on the use of environmental design software among architects showed that almost none of the architects surveyed employed such tools in their practice, and that consultations with building scientists are rare (Wong et al., 1999).

The second thread of research, typified by the Olgyays' bioclimatic approach to design, found more followers and inspired later generations of architects like Ken Yeang and Tay Kheng Soon. Olgyay and Olgyay (1963), in *Design with Climate: Bioclimatic Approach to Architectural Regionalism*, stated that architects do not have the time and mental resource to compute all the bioclimatic data. They propose that calculations on how parametric variations in building form can affect indoor human comfort in a climatic context be made separate from the architect's design process, to produce principles and graphical guidelines for the architects to use as a generator towards a regionalized architectural language.

Over the years, these parametric studies have grown and have become more sophisticated with attempts at integrating the different fields in building science and addressing other environmental and sustainable issues.

1.4.1   Social and cultural dimensions

One challenge for building science is the influence of social and cultural dimensions in the application of building studies into architectural design. Over the years, social and cultural rituals have evolved to adapt to climatic conditions and vernacular architecture, in particular, embodies some of these adaptations in their plan and design features (Hawkes, 1996). It is important that these socio-cultural factors are somehow subsumed or incorporated into building science studies. While socio-cultural factors are a common thread in the writings of many architects, such concerns have not found a suitable quantitative expression. Various essays in this book observe the sensibility and necessity of such integration, and suggest ways to think about and correlate these seemingly opposed dimensions.

In "Rethinking Design Methodology for Sustainable Social Quality" (Chapter 2 of this book), Alexander Tzonis observes that an emerging movement is drawing attention to the problem of "sustainable" artificial and natural environment, initially related to its physical quality, but increasingly associated with social quality and resulting in new knowledge. He suggests that methods and architecture theory do not end like movements that rise and fall and points out that Design Methods and Critical Theory, thought to have ended, are tools that can enrich design thinking and understanding in architecture and urban planning, and especially in this current direction.

In "The Selective Environment: Environmental Design and Cultural Identity" (Chapter 3), Dean Hawkes provides another perspective about architecture as it relates to environmental response and cultural identity. He suggests that "The Selective Environment is an approach to environmentally responsive architectural design that seeks to make connections between the technical preoccupations of architectural science and the necessity, never more urgent than today, to sustain cultural identity in the face of rapid global technological change." He discusses how selective theory facilitates the connection between historical analysis in architecture and contemporary practice and building technology, with particular relations to comfort and climate, nature and architecture and the environment and regionalism.

1.4.2   Urban issues

Another great concern is the effect of high-rise high-density buildings and overcrowding in the rapidly expanding city. While the air-conditioned high-rise is easily replicated in the tropical city, the effects of urban canyons and heat entrapment in the city are different for the tropics. While sunlight is welcome in the temperate city and buildings are set back to allow sunlight to penetrate to the road level, shade is preferred in the tropics. While snow and sleet may be a problem in temperate cities, the problem in the tropics is heavy rain and flooding. While strong gales are better avoided in colder cities, more wind and ventilation are welcome in the tropical (and sub-tropical) city. It is only recently that urban studies have been made to some depth in tropical cities and the findings are suggestive in terms of the design of the tropical city for the future.

In "The Ecological Design of Large Buildings and Sites in the Tropics" (Chapter 4), Ken Yeang suggests an ideal for relating all built works ecologically to nature and provide a practical framework by which one can deal with environmental issues of large developments within the dense urban context. He states that "... the ecological design ideal involves the holistic and careful consideration of the use of materials and energy in built systems and the endeavour, by design, to reduce their undesirable impact on, and to integrate them with, the natural systems of the locality over their entire life-cycle."

There are several other selected essays in this book that cast light on various environmental issues, policies and guidelines for the dense and rapidly transforming tropical city, as well as essays on experimental sustainable projects relating to community and old city fabrics, and studies on interrelationships of the social and environmental conditions for quality of living and towards a more holistic sustainability. These are introduced briefly below.

1.5 SELECTED ESSAYS

1.5.1 Socio-environmental dimensions in high-rise high-density living context

Is it possible to have a modern high-rise vernacular architecture? Can design guidelines for the architect go beyond the bioclimatic approach to include social dimensions? In "Socio-Environmental Dimensions: In Tropical Semi-open Spaces of High-rise Housing in Singapore" (Chapter 5), Joo-Hwa Bay, Na Wang, Qian Liang and Ping Kong show an example of a high-rise kampong (village) in Singapore, and discuss a possible framework to understand the inter-relationships of the socio-environmental dimensions in the tropical semi-open living space. They also propose a set of guidelines for future design that embodies the socio-climatic potentials, suggesting a step beyond the bioclimatic, presented graphically and structured for the qualitative and heuristic design thinking of the architect.

Are the current assessment methods for sustainable built environment adequate? Are they limited to building science criteria, lacking consideration for socio-economic aspects relating to architecture? In "Building Environmental Assessment Methods from Sustainable Architecture Perspective: An Analysis in the Singapore Public Housing Context" (Chapter 6), Boon-Lay Ong and Chi-Nguyen Cam show how socio-economic dimensions can contribute not only towards a more sustainable quality of living, but also to quantitative environmental performances. They propose that socio-economic criteria should be developed and integrated with environmental criteria in the assessments of sustainable housing developments.

1.5.2   Policies, building and planning guidelines

Another criticism of current sustainable assessment methods is that they are mostly developed for the colder climatic zones. In "Policy and Evaluation System for Green Building in Subtropical Taiwan" (Chapter 7), Hsien-Te Lin discusses an assessment system that is applicable to the Asian tropical and sub-tropical context. He explains a green building evaluation system with four evaluation categories, Ecology, Energy saving, Waste reduction and Health (EEWH), and nine other environmental indicators, simplified and localized for Taiwan. This system has been regarded as a standard evaluation method for Green Buildings by the Ministry of the Interior of Taiwan since 1999, as a framework for Green Promotion programs and as a mandatory Green Building policy.

The EEWH system furnishes the outdated building and planning guidelines with relevant criteria towards more sustainable development of the city. In Singapore, a similar system has been developed called Green Marks, but it is yet to be made mandatory and integrated with the building and planning guidelines.

Q.M. Mahtab-uz-Zaman, Jalal Ahmad, Fuad H. Mallick and A.Q.M. Abdullah attempt to evaluate the inadequacy of existing building and planning guidelines with environmental criteria so as to propose a new set of guidelines that integrates sustainable qualities in Dhaka. They discuss a proposal for a new set of Floor Area Ratio (FAR) guidelines in "In Search of a Habitable Urban Space-Built Ratio: A Case Study of Building and Planning Regulation in Dhaka City" (Chapter 8). This new FAR guidelines seek to ensure comfortable indoor and outdoor environment; create more green areas to reduce urban heat island effect, as well as more outdoor space for social activities; create a better ecological balance and preserve low lying areas for water retention as flood protection mechanism – all of which collectively has the potentials to enhance social and environmental qualities of the city.

1.5.3   Urban environmental impacts

In the tropical regions, it is important to provide natural outdoor conditions that are conducive to social activities in the urban context, in terms of adequate daylighting, shading from solar radiation and ample ventilation. A cooler and brighter external environment will also contribute to the quality of the interior environment of individual buildings. In "Designing High Density Cities: Parametric Studies of Urban Morphologies and Their Implied Environmental Performance" (Chapter 9) Edward Ng, Tak-Yan Chan, Vicky Cheng, Nyuk Hien Wong and Meiqi Han discuss how varying the skylines of tall buildings in the very dense-built environment of Singapore and Hong Kong can improve performances of daylighting and natural ventilation in the outdoor spaces. The performances are significantly higher as compared to having buildings with uniform height, and warrant some rethinking of the typical simple height zoning in master planning.

Excessive heat in the urban environment can have serious negative impacts on urban dwellers resulting in heat-stress and higher energy usage because of the need for air-conditioning. Outdoor activities can be uncomfortable, and air-conditioning can further increase the heat quantum, escalating the problem, as well as increasing pollution. Heat sources in the urban context includes heat re-radiated from building materials exposed to the sun, heat generated from combustion processes, air-conditioning, and greenhouse effect owing to pollutants in the atmosphere. In "Exploring the Urban Heat Island Effect in Singapore" (Chapter 10), Nyuk-Hien Wong and Yu Chen discuss their research on the severity of Urban Heat Island (UHI) effect on the urban environment of Singapore, and offers suggestion for mitigating some of the impacts.

In "Thermal Environment Study of Urban Canyons" (Chapter 11), Elias Salleh observes that urban outdoor spaces (urban canyons) in between buildings of differing heights exhibit different microclimatic conditions that can affect the conduciveness for social activities. Generally, a deeper urban canyon provides more shade from solar radiation and is comfortable with lower wind speeds, compared to a shallower canyon. He offers several guidelines to designing urban canyons, by studying empirical examples in Kuala Lumpur, Malaysia. His study can also inform the environmental assessments of existing city fabrics, especially the older city fabrics, for various planning decisions for conservation and renewal.

1.5.4 Experimental sustainable projects

Often old city fabrics exist in many cities, requiring decisions of reuse or redevelopment owing to environmental problem heat and pollution owing to intensification of urban activity and commercial pressures. In "Tropical and Traditional: Inventing a New Housing Model for the Old 36 Street Quarter in Hanoi, Vietnam" (Chapter 12), Shoichi Ota discusses how conducive environmental quality can be achieved for a very long and deep lot in the old city fabric with a modern experimental infill residential structure. He discusses the concept of "space block" by the architect, Kazuhiro Kojima, where the modern block structures are composed for maximum porosity, and shows the level of success of this project with an environmental assessment of the thermal comfort level through maximum ventilation and shading. This experimental example offers suggestions and hope for architects and planners for the reuse of historical city fabrics such as this for new uses with confidence that there will be no slack in the environmental quality.

José Roberto Garcia Chavez discusses a different experimental dwelling project in "ECOPET 21: An Innovative Sustainable Building System for Ecological Communities in Tropical Regions" (Chapter 13). He explains the benefits, principles and the whole process of an innovative sustainable construction system called ECOPET 21 for a prototype housing community, integrated with the application of bioclimatic design principles, sustainable technologies of renewable energies, and environmental planning. Much can be learned from this experimental project for thinking about future sustainable community housing developments.

1.6 TOWARDS TROPICAL SUSTAINABLE ARCHITECTURE

The essays here define various dimensions and present the latest research and thinking that surrounds the field of tropical architecture today. However, the issues presented here are relevant to the architectural discourse elsewhere in the world also. Frank Lloyd Wright is known to have said that his buildings are not of any style but simply designed with style. Style in this sense is not a visual appearance but a consequence of a particular way of looking at design or of the design process.

It seems logical to expect that the approach to architecture presented here will naturally lead to a tropical sustainable architecture, but however the problematic of tropical architecture language discussed in the introduction still lingers on and continues to be debated. We end this book with a philosophical discussion in the last chapter, presented by Anoma Pieris in "Is Sustainability Sustainable? Interrogating the Tropical Paradigm in Asian Architecture".

We hope that this book will spur deeper reflections and more research into the issues of tropical sustainable architecture that engages both the environmental and social dimensions.

NOTES

1 iNTA aims at promoting international research and collaboration on studies relating to sustainable architecture and urban design, relating both the social and the environmental dimensions, in the

tropical and sub-tropical regions. The iNTA website is currently located at: http://www.arch.nus.edu.sg/inta/index.htm

2 The essays in this book are different from the related papers in the published proceedings of iNTA 2004 Conference hosted by the National University of Singapore.

3 These two architects, as do many other architects in Singapore and Malaysia, work in both countries because of their close physical, political, social and cultural ties.

REFERENCES

Banham, R. (1984) *The Architecture of the Well-tempered Environment* (2nd ed.) Chicago: The University of Chicago Press. Originally published in 1969.

Bay, J.H. (2001a) Three tropical design paradigms, in Tzonis, A., Lefaivre, L. and Stagno, B. (eds), *Tropical Architecture: Critical regionalism in the Age of Globalisation*, London: Wiley-Academy, pp. 229–265.

Bay, J.H. (2001b) Cognitive Biases in Design: The Case of Tropical Architecture. Ph.D. dissertation, The Netherlands: Design Knowledge System, TUDelft.

Bess, P. (1996) Communitarianism and Emotivism: Two Rival Views of Ethics and Architecture, in Nesbitt, K. (ed.), *Theorizing a New Agenda for Architecture*. Princeton: Princeton Architectural Press.

Burckhardt, L. (1992) On ecological architecture: A memo, in Tzonis, A. and Lefaivre, L. (eds), *Architecture in Europe: Memory and Invention since 1968*, London: Thames and Hudson, pp. 42–43.

Fry, M. and Jane, D. (1956) *Tropical Architecture in the Humid Zone*. London: Batsford.

Fry, M. and Jane, D. (1964) *Tropical Architecture in the Dry and Humid Zone*. London: Batsford.

Hagan, S. (2001) *Taking Shape: A New Contract between Architecture and Nature*. Architectural Press: Oxford.

Hawkes, D. (1996) *The Environmental Tradition: Studies in the Architecture of Environment*. London: E & FN Spon.

Hyde, R. (2000) *Climate Responsive Design: A Study of Buildings in Moderate and Hot Humid Climates*, London and New York: E & FN Spon.

Jacobs, J. (1962) *The Death and Life of Great American Cities*. New York: Random House.

Lefaivre, L. and Tzonis, A. (2001) The Suppression and Rethinking of Regionalism and Tropicalism after 1945, in Tzonis, A., Lefaivre, L.

and Stagno, B. (eds), *Tropical Architecture: Critical Regionalism in the Age of Globalization.* Chichester: Wiley Academy.

McDonough, W. (1996) Design, Ecology, Ethics, and the Making of Things, in Nesbitt, K. (ed.), *Theorizing a New Agenda for Architecture.* Princeton: Princeton Architectural Press.

Mumford, L. (1961) *A City in History: Its Origins, its Transformations, and its Prospects.* Paperback reprint, 1991. Penguin.

Olgyay, V. and Olgyay, A. (1963) *Design with Climate: Bioclimatic Approach to Architectural Regionalism.* Princeton: Princeton University Press.

Olgyay, V. (1952) Bioclimatic Approach to Architecture, in The Building Research Advisory Board, 1953, (ed.), *Housing and Building in Hot-humid and Hot-dry Climates.* Research conference report No. 5, Washington, D.C., Building Research Advisory Board, pp. 13–23.

Ong, B.L. and Hawkes, D.U. (1997) The Sense of Beauty: The role of Aesthetics in Environmental Control, in D. Clements-Croome, (ed.), *Naturally Ventilated Buildings: Building for the Senses, the Economy and Society,* E & FN Spon: Chapman & Hall, pp. 1–16.

Ong, B.L. (1997) From Homogeneity to Heterogeneity, in D. Clements-Croome, (ed.), *Naturally Ventilated Buildings: Building for the Senses, the Economy and Society,* E & FN Spon, Chapman & Hall, pp. 17–34.

Ong, B.L. (1994) Designing for the Individual: A Radical Interpretation of ISO 7730, in Humphrey, M., Sykes, O., Roaf, S. and Nicol, F. (eds), *Standards for Thermal Comfort.* London, Chapman & Hall, pp. 70–77.

Pallasmaa, J. (1993) From Metaphorical to Ecological Functionalism, *The Architectural Review*, 193, June.

Pallasmaa, J. (1996) *The Eyes of the Skin: Architecture and the Senses.* London: Academy Editions.

Pérez-Gómez, A. (2005) Ethics and Poetics in Architectural Education. *Folio 06: Documents of NUS Architecture.* Singapore: National University of Singapore.

Tzonis, A. and Lefaivre, L. (2001) Tropical Critical Regionalism, in Tzonis, A., Lefaivre, L. and Stagno, B. (eds), *Tropical Architecture: Critical Regionalism in the Age of Globalisation*, Prince Claus Fund for Culture and Development, The Netherlands. London, Wiley-Academy, pp. 1–13.

Tzonis, A. and Lefaivre, L. (1990) Why Critical Regionalism Today? in Nesbitt, K. (ed.), *Theorizing a New Agenda for Architecture: An Anthology of Architectural Theory 1965–1995*, New York: Princeton Architectural Press.

Tzonis, A., Lefaivre, L. and Stagno, B. (eds), *Tropical Architecture: Critical Regionalism in the Age of Globalisation*. London: Wiley-Academy.

Tzonis, A. (1972) Towards a Non-oppressive Environment. Boston: Boston Press.

Watson, D. (1991) Commentary: Environmental Architecture. *Progressive Architecture* 3.91.

Watson, D. (1995) Sustainability: The roots and fruits of a design paradigm. In *ACSA Proceedings 83rd Annual Meeting*, (March 1995). Seattle, Washington.

Wong, N.H., Lam, K.P. and Feriadi, H. (1999) The Use of Performance-based Simulation Tools for Building Design and Evaluation: A Singapore Perspective. *Building and Environment*. Vol. 35, Great Britain: Elsevier Science Ltd., pp. 709–736.

Part I

ARCHITECTURAL AND ENVIRONMENTAL THEORIES

2 RETHINKING DESIGN METHODOLOGY FOR SUSTAINABLE SOCIAL QUALITY

Alexander Tzonis

Technological University of Delft

Abstract

Half a century ago, the Design Methods movement was acclaimed to be a science-based panacea for all architectural and planning problems. By the middle of the 1970s it had plunged into a period of neglect. Many declared the end of "design methodology" as a discipline, and the dominant place that Design Methods had occupied as a movement was taken over by an opposite approach to architecture and urban planning known as "Critical Theory." Today, as Critical Theory is also passing fast into oblivion, many ex-advocates of this discipline claim that Architectural Theory has also ended. Of course neither Design Methods nor Architectural Theory are coming to an end, at least in the foreseeable future. Only movements rise and fall ("theory" or "method" being only labels of these movements) shifting attention and priorities that reduce or enrich architectural thinking. Such is the case of the current movement that appears to be emerging and drawing our attention to the problems of "**sustainable**" environment – whether generated by humans or naturally. Initially related to its physical quality it is increasingly associated with **social quality** and its competence to create new knowledge (Tzonis, A. The Creative City Shenkar College Seminar on the City, 2005).

Keywords

Architectural Thinking, Design Methods, Critical Theory, Regionalism, Creativity, Sustainability.

2.1 INTRODUCTION

Concern with social quality was not alien to architectural movements of the past and certainly not to either "Design Methods" or "Critical Theory." What is unique about the current design movement (which we might call Social Sustainability) is (a) its pragmatic outlook and its concern with tangible environmental results, and (b) its apprehension about social quality of our buildings and cities in the long run.

Characteristic of the 1960s Design Methods movement was its extreme trust in positive sciences, empirical and analytical, and its constructive and overoptimistic disposition. Even if the conclusions of Design Methods papers were highly abstract, their declared intention was to reform and improve design practice. By contrast, the Critical Theory movement which followed a generation later was indifferent, if not hostile, to the scientific approach and skepticism, if not pessimistic, about its helpfulness.

Why this shift? What went wrong with the Design Methods movement? Why did the idea of transferring methods from natural sciences to design become so detested? Why was the idea of being disinterested and having nothing to do with practical involvement seen with such positive eyes?

There are many reasons: first, by the end of the 1960s a large number of researchers showed that there were strong links between the introduction of the way of thinking from the natural sciences to architecture and the growing needs of the Welfare State for economic efficiency, technical effectiveness, and social control. Researchers found that this development had taken place gradually since the seventeenth century, accelerated during the twentieth century, and arrived at a peak immediately after the Second World War when, as Vannevar Bush stated in his famous memorandum to President Roosevelt, science was expected to solve the problems of post war society as it had already done prior to the problems of war (Bush, 1945; Hughes, 1989).

In the specific case of architecture and urban design, Design Methods (transferring "methods," ways of inquiry, from sciences to design) was supposed to solve problems related to housing, urban growth, and transportation which traditional practice could not solve because it lacked methodology, rigorous procedures for collecting and organizing data, and principles for rational decision-making and problem-solving techniques. As several of the techniques that also originated from Design Methods were introduced into professional practice, it was shown that they did not work either. Thus, the claim that the Design Methods movement could reform architecture and urban design was shaken.

But there were also political objections. There were accusations that this transfer was motivated by material interests rather than pure intellectual pursuits. The basic conceptual framework of the Design Methods movement, it was argued, was constructed to serve the grand top-down planning strategy of the Welfare State architecture as this materialized after the First World War. In addition, apart from the fact that this dependence made the Design Method movement politically suspect – at least to some populist leaning groups – it also made it vulnerable to the criticism that it was becoming increasingly irrelevant. This was because the fiscal crisis of the public institutions and international economic development was bringing about the rapid decline of the Welfare State.

Finally, many theoretical doubts were raised, questioning the legitimacy of the transfer of a natural science ways of inquiry into the world of social and cultural objects such as buildings and cities. The author has argued that much of what passed as "scientific design" in the 1950s and 1960s was, in fact, only "scientistic," based on shallow reading of similarities of data and superficial analogies. In most cases the Design Methods approach to design had the external characteristics of science but not its deeper logic, the error originating out of the fact that the Design Methods movement lacked theoretical foundations and a reflective critical outlook.

2.2   CRITICAL THEORY

Critical Theory derived its name from Kant's term "critical." Kant was the originator of "critique" as a philosophy program, a program with roots in Socrates, which established as its task "a science of the mere examination" of "the sources and limits" of "pure reason," of the *a priori* "principles whereby we know anything." In contrast to Design Methods, Critical Theory declared that it had no intention to reform the architectural profession.

Kant defined the "utility" of this inquiry as "negative," the concept of "negative" meaning here that the "critique" was "not to extend, but only to clarify our reason, and keep it free from errors" (Kant, 1965). Recasting this idea in a post-Marxist frame Critical Theory was expected to inquire into the "sources and limits" of epistemology or to be closer to the case, of "dominant ideology." In contrast to Design Methods which produced several strategies for reforming architectural practice, Critical Theory developed only a vague notion of "resistance," a "negative" (not in the Kantian sense) notion implying refusal to go along with professional practice and becoming the handmaid of the design profession even if this would have contributed to environmental amelioration.

Within the circumstances of the decline of the Welfare State, the idea of a "critical" ("negative" theory, especially as it was injected by not only Marxist but also Nietzschean, Heideggerian, and Freudian ideas) became part of the popular attacks against its professional institutions and their theoretical presuppositions on which they were grounded. At best, these assaults tried to show that historically designing as thinking and as action were driven by highly private intentions, power struggles, and collective memories and, as such, they could not be dealt with by transferring logico-empirical methods applied in mechanics. Accordingly – and one might say to some extent not "critically" enough – proponents of Critical Theory rushed to declare design situations as intractable and "wicked," to use the expression of Horst Rittel (1973), thus having no use of reductive or idealized models or problem-solving processes which were developed by Design Methods. In particular, even before Critical Theory became fashionable, Rittel and his Berkeley team declared (in the mid-1960s) that it was impossible to solve the majority of urban policy or planning problems within the framework of any traditional problem-solving techniques of that time, pointing to a convincing number of concrete cases. The vacuum created by the expulsion of scientific models and techniques from architecture Critical Theory asked to be filled in through further "critical" discussions about ideology and politically interpreted epistemology, regardless of their relevance or effectiveness in improving environmental conditions.

2.3   DESIGN METHODS

Design Methods vanished from the curricula of most Departments of Architecture, and most columns of architectural journalism, having achieved a very modest influence in the profession by introducing a small number of techniques for improving the efficiency and effectiveness of practice. Similarly, the impact of Critical Theory – preoccupied with the "limits" of design rather than its potentials – was felt mostly in academic debates in areas such as schools of design, architectural journalism, and at architectural cultural events. Like Design Methods, the influence of Critical Theory on architectural firms was minimal. It did manage, however, to curtail the intrusion of Design Methods in practice, probably saving many firms from unnecessary costs and even possible errors. It also succeeded in driving away technocratic projects promoted through pseudo-scientific arguments where their desirability was doubtful and their environmental and social impact not given sufficient thought. Perhaps its most important influence was to make architects, clients, and the public more conscious of history, creating a more positive climate for historical preservation and a greater awareness of issues of cultural meaning of sites.

2.4   THE BUILDING BOOM

Never before in history there was a building boom such as the one that followed the end of the Second World War. The primary motor for this massive construction was initially the Welfare State (notably in the United Kingdom) and, by the end of the Cold War, private developers. While many of these projects were seen as satisfactory in the *short-term*, in the *long-term* a very large number of them were recognized as failures. What happened was that where major projects of the second half of the twentieth century achieved instant success through mass-publicity, a few years after their construction they proved to be unworkable, aging fast and, as a result, generating an environment of embarrassingly bad poor quality, physical and social areas. The building boom was followed by a boom of massive demolitions. As a popular Television program *Demolition* showed, news of buildings being demolished were often received by the public more enthusiastically than the news of their construction (The Guardian, 2005). This was not so much the result of shifting fashions as the fact that these projects did not deliver an environment of *sustainable* quality. As a result of an accumulation of such failures, by the mid-1990s, design *sustainability* emerged as the new leading movement.

2.5   THE SUSTAINABLE DESIGN MOVEMENT

It was not in architecture that the term *sustainability* was introduced for the first time[1]. Already by the end of the 1980s, the term "sustainability" was used extensively in the economics field in reference to development for criticizing earlier models of economic growth for nations or regions that had favored fast returns and accelerated growth, while disregarding that in the long-run they were depleting irreplaceable resources – the very resources their growth depended upon.

The question of the long-term, unanticipated negative impact of an economic policy on its performance was further extended to cover the effect of new products – chemical, agricultural, and mechanical – on environmental quality in the long run. It was in relation to this latter problem that the criterion of sustainability entered into architecture and urban design, providing a conceptual framework to handle the long-term negative impact of the application of techniques and materials of construction on material resource consumption and environmental physical quality.

Many of these issues were discussed in the 1960s as pioneering research as publications from Donald Watson and the writings of Lucius Burkchardt (Tzonis et al., 1992) demonstrate. And one should never forget the ground-breaking writings of Lewis Mumford. In relation to these earlier efforts, the current Sustainable Design movement is more analytical, critical, and theoretical, which clearly shows that both the Design Methods and the Critical Theory approaches had, after all, some influence on its development.

2.6   SOCIAL QUALITY

The question of sustainability of the *social* quality of the environment was also addressed by the end of the 1950s. Most influential were the writings of Jane Jacobs. Jacobs was neither an architect nor an urban system's engineer but it was her articles and later her books that made people aware that the built environment was not a static artifact but a container of users and uses whose performance evolved in a complex, dynamic, and interdependent manner, the path of the evolution constrained by the artifact's physical attributes. Rather than focusing on built form as a thing in itself – as most architects and urban planners did until the beginning of the 1960s – Jane Jacobs looked at the urban tissue as a network of conduits that have an impact on who relates to whom and how. By doing so, Jacobs succeeded in explaining why the social quality of some parts of a city thrive over time while others decline.

Jane Jacobs identified social quality of a place in terms of the quality of human interactions within it. Thus, from the point of view of sustaining this quality, it is people in relation to people which are the resources that must not be depleted. To be more exact, what has to be sustained is the potential of contact and encountering people. In this case "people" is not a question of number. Quality depends on the relevance of a person in relation to another. Relevance in turn can be defined in terms of short-term targets: instant services for immediate needs. On the other hand, it can also be defined in reference to the long-term potential of people to interact with each other in order to understand new phenomena, to create new knowledge, and to continue adapting to new challenges. Indeed, it has been shown in general that the *diversity* of knowledge resources contributes to knowledge creation.

2.7   ENVIRONMENTAL DIVERSITY

Why diversity? In ecology, the diversity of animal and the diversity of plant communities are closely interlinked. It is a major factor contributing to survival and results in species-variety-poor or species-variety-rich environments; for example, in areas which are dominated by one or few related species (like northern forests) or, conversely, in areas which have an infinite variety like tropical forests. Contemporary sociological and economic case-studies and historical research have indicated that human-made environments that have enhanced potential interactions between people of high cultural, social, and disciplinarian diversity have sustained the generation of new knowledge and creativity. In turn, new knowledge and creativity helped sustain a social-cultural quality and community within a world of changing fortunes. In other words, there are good reasons to believe that there is a strong relationship between social sustainability and diversity in potential human interactions.

2.8 KNOWLEDGE RESOURCES

Today however we have journals, books and, most importantly, the web that makes knowledge resources available to all. Is physical access of people to people still needed then?

Empirical studies suggest they do: as Edward Glaeser (2004) has stated, "proximity to the coal mines or the harbor may have mattered in 1900, but do not matter today. Instead, the productive advantage that one area has over another is driven mostly by the people." In other words, researchers have confirmed that people need to have physical accessibility to knowledge resources as these are "carried around" by people rather than contained in texts or digital data bases. Further, an interesting conclusion reported by researchers is that "virtual" interaction between knowledge resource people that is electronic-based (especially the web) does have disadvantages. It leads to a phenomenon called Cyber-balkanization (Van Alstyne et al., 1996); that is, the clusters formed by interacting people through an electronic media tend to become less diverse as time goes by. To put this in a different way, the exclusive reliance on electronic media generates non-sustainable environments from the point of view of diversity, and consequently less conducive to "idea creation." That inevitably has consequences in planning, and considerable studies have been carried out on location choice, transportation development, and the formation of "clusters" maximizing the potential of "human capital" (Porter, 1998).

2.9 SUSTAINABILITY AND SOCIAL QUALITY

A few, but important contributions, have been made by architects who tried to conceive their projects in terms of human interaction.

However, these projects related to the problem iconically, and the fact that the urban and building fabric work can be seen in terms of nested networks of conduits through which people come together face-to-face and interact, has not been developed (Wu, 2005). The idea of sustainability of social quality was not explicitly stated in their projects, although in their writings, and the justification of their projects, they stressed the issue of "time," "urbanity," and "urban community" (Woods, 1972).

Louis Kahn developed a number of diagrammatic and built manifestoes, the most significant being the Plans for the Center of Philadelphia, the Bath House in Trenton New Jersey, and the Richard Laboratories in Philadelphia. The manifestoes invited people to look at buildings and cities as manifestations of movement organized in terms of "rivers and docks," "serving and served" components enabling human contact. Similarly, about the same period, Aldo van Eyck, a Dutch architect, built a series of experimental schemes, the most important being his 700 Amsterdam Playgrounds in the 1940s and 1950s, the Municipal Orphanage, for the City of Amsterdam (1954), and the Sculpture Pavilion for Sonsbeek Exhibition, Arnhem (1965–66). Through these he urged others to look at the "voids," the "in between" buildings as spaces where people meet, "places" for "events," and for human "encounter," rather than as abstract volumes.

At the end of the 1950s, Shadrach Woods, an ex-collaborator of le Corbusier generated polemics hosted in the pages of a very small in size, but influential magazine, *Le Carré Bleu.* These articles criticized the "obsolete formalism of monumental architecture," and the "plastic or aesthetic arrangement . . ." of Modernist Architecture of the post war era. "The problems which we face in making our world are entirely new, for our society is entirely new," he stated. ". . . a completely open, non-hierarchical co-operative in which we all share on a basis of total participation and complete confidence . . . We cannot think of planning in static terms, in three-dimensional space, when we live in a four-dimensional world . . . the scene of action of reality is *not a three-dimensional Euclidean space but rather a four-dimensional world* [my emphasis] in which space and time are linked together indissolubly." He thus put forth the analytical design concepts of "stem," "cell," and the "web" to focus on buildings and cities as tools of *social interaction* and *maximization* of *choice*. Woods tried to implement these ideas in the scheme for the Center of Frankfurt, which was not built at the time, and for the Free University of Berlin which has since been remodelled and still passionately debated.

Similar ideas generated Yona Freedman's 3D networks, and in the academic world, Serge Chermayeff's Yale influential multidisciplinary theoretical investigations on social interactions and built form, known as the "shape of community" (Serge and Tzonis, 1971).

What all these architects tried to do was to develop a new kind of architecture reacting to the 1950s Melvin Webber's idea of achieving a community without the physical proximity of place, Karl Deutsch's "city as a switchboard," and the renowned "Global Village" of Marshal McLuhan envisaging an empire of global communication without any human contact.

However, none of these efforts have produced tangible answers on how to design buildings and cities from the point of view of social quality which would enable diversity of human interaction.

2.10 DESIGN STRATEGIES

There are many avenues of investigation open for constructing new design strategies for sustainable social quality environments. As in designing sustainable ecological environments, we have begun to understand how the choice of materials, orientation, dimensioning and proportioning of spaces, of solids and voids, etc. affect long-term physical quality. So in designing for sustainable social quality we have to explore and discover how decisions about the spatial structure of the environment as a communicator enable interactions. The field to explore is enormous and the task of inquiry is just beginning.

Before my conclusion, I would like to suggest a warning on how to avoid the trap of reductionism and the bias of environmental determinism which architects are very prone to when dealing with complex problems for which no easy rules exist (Bay, 2001).

We have discussed what makes buildings and cities sustain social quality. We have identified the diversity of people interacting with each other as one of the contributing factors towards social quality and focused on the problems of how to design towards this desirable state. However, to bring about diversity certainly presupposes the existence of groups that are different from each other. Thus, a reciprocal design problem emerges: how to design environments which enable individuality, identity and, consequently, difference. This is a problem we have not discussed here, although it is a fundamental component if designing towards sustainable social quality environments has to work. Even when people meet this does not imply that they would talk to each other about generating social quality. Given differences in education acquired, institutional conventions, and certainly interests, the probability is that they would be incompatible or incommensurable with each other's differences in beliefs, knowledge, and understanding. The fact that a project provides the opportunity for people to meet, does not lead automatically to agreement and collaboration. Quite the opposite – it might make conflict possible. We should not forget, therefore, that designers provide the *necessary* physical conditions for things to happen. It enables but does not have the power, or the professional responsibility, to cause things to happen. This rests beyond architecture.

I would suggest two directions towards such necessary design conditions for enabling sustainable social quality:

1. An investigation into the design of environmental structures that on one hand maintain differences in identities and on the other provide for *"speciation,"* new "species" of cultures and new social identities (what E.J. Mishan, the great pioneer of the sustainability movement, once called, *separate facilities*).
2. An investigation into the design of environmental structures that can aid *accessibility* between people of maximum diversity.

As in ecology, so in designed environments, sustained diversity becomes manifest in luxuriant settings; however, diversity is not a luxury. It is one of the most significant conditions needed for long-term survival of life as a biological and as a cultural phenomenon.

NOTE

1 *Sustainable development* first became well-known after the publication of the *World Conservation Strategy*, published by the World Conservation Union (IUCN) in 1980. The IUCN subsequently achieved a new status with the publication of *Our Common Future*, the Brundtland Report, in 1987 and has obtained even greater attention since the United Nations Conference on Environment and Development (UNCED) held in Rio de Janeiro in June 1992. Now the concept involves governments and non-government organizations (NGOs), civil servants and environmental activists, and local government. "Sustainable" is derived from the Latin verb "sustinere" and describes relations (processes or states) that can be maintained for a very long time or indefinitely (Judes, 1996).

REFERENCES

Bay, J.-H. (2001) Cognitive Biases in Design: The Case of Tropical Architecture. The Netherlands. TUDelft: Design Knowledge System Research Centre.

Bruntland, G. (1987) *Our Common Future: The World Commission on Environment and Development*, The Bruntland Report, Oxford University Press.

Bush, V. (1945) *Science the Endless Frontier*.

Glaeser, E. (2004) Review of Richard Florida's *The Rise of the Creative Class*, Papers on the Web, Havard University, May.

Hughes, T.P. (1989) *American Genesis: A Century of Invention and Tecnological Enthusiasm*, New York.

IUCN, *The World Conservation Strategy*, IUCN, UNEP, WWF, 1980.

Kant, I. (1965) *Critique of Pure Reason* Trans. New York: Norman Smith.

Porter, M.E. (1998) Clusters and the New Economics of Competition. *Harvard Business Review*, 76, no 6.

Rittel, H.W.J. and Webber, M.M. (1973) Dilemmas of a General Theory of Planning, *Policy Sciences* 4, pp. 155–169.

Serge, C. and Tzonis, A. (1971) *Shape of Community*. London: Penguin Books.

The Guardian (2005) *Watch it Come Down*, August 13.

Tzonis, A., Cohen, J. and Lefaivre L. (1992) *Architecture in Europe*, Rizzoli.

Tzonis, A. and Lefaivre, L. (1952) *Planning and Tomatoes*, Casabella, Jan–Feb, pp. 146–149.

Van Alstyne, Marshal and Brynjolfsson, E. (1996) Wider Access and Narrower Focus *Science* 274, 5292.

Woods, S. (1972) *Man on the Street*. London: Penguin Books.

Wu, J. (2005) *A Tool for the Design of Facilities for the Sustainable Production of Knowledge*. Delft: Design Knowledge Systems.

3 THE SELECTIVE ENVIRONMENT: ENVIRONMENTAL DESIGN AND CULTURAL IDENTITY

Dean Hawkes

*Welsh School of Architecture, Cardiff University, UK
and Darwin College, University of Cambridge, UK*

Abstract

The Selective Environment is an approach to environmentally responsive architectural design that seeks to make connections between the technical preoccupations of architectural science and the necessity, never more urgent than today, to sustain cultural identity in the face of rapid global technological change. This paper summarizes the fundamentals of selective theory, with particular emphasis on the themes of the *environment and regionalism, nature and architecture,* and the relation of *comfort and climate.* The method proposes the connection of historical analysis in architecture with contemporary practice and building technology.

Keywords

Architecture, climate, comfort, environment, nature.

3.1 INTRODUCTION

"... the main antagonist of rooted culture is the ubiquitous air-conditioner. Wherever they occur, the fixed window and the air-conditioner are mutually indicative of domination by universal technique."

In this statement, Kenneth Frampton (1983) identifies the air-conditioner as a potent symbol of the way in which the environmental function of architecture was transformed during the later decades of the twentieth century. Through the agency of mechanical systems, operating within sealed building envelopes, Le Corbusier's proposition of, "... only one house for all countries, the house of *exact breathing*" has become globally commonplace (Le Corbusier, 1930).

In contrast to this powerful stereotype we have seen the emergence of a strong, alternative line of thought. This rests upon the case for a *regionally* grounded approach to contemporary design. In its environmental manifestation, the regionalist position was probably first argued by Victor Olgyay, in *Design with Climate: Bioclimatic Approach to Architectural Regionalism* (1963). More recent and continuing advocacy for regionalism, in a wide interpretation, has come from Alexander Tzonis and Lianne Lefaivre in a sequence of important publications (Tzonis and Lefaivre, 2001). The idea of *The Selective Environment: An Approach to Environmentally Responsive Architecture* (Hawkes et al., 2002) is located within the regionalist paradigm and proposes a broad strategy for design that embraces the technical and the cultural in the realization of a sustainable architecture.

3.2   DEFINITIONS

In *The Architecture of the Well-tempered Environment*, Reyner Banham (1969) proposed three distinct "modes" of environmental control that may be applied in architecture. He named these the "Conservative", the "Selective" and the "Regenerative". Banham's classification was derived from empirical observation of historical building types and effectively served the ends of his primarily historical analysis. In 1980, the present author adapted Banham's categories in order to make a clear distinction in environmental design strategy that could be observed in contemporary design practice. This defined two modes of environmental control, the "Exclusive" and the "Selective," the principal characteristics of which are summarized in Table 3.1.

These modes correspond precisely to distinction between "global" (exclusive) and "regional" (selective). In a later representation it was felt necessary to admit to the existence of a vast and uncharted body of buildings in which environmental design is, regrettably, more a matter of chance than of design. This was represented in an adaptation of Ebenezer Howard's "three magnets" diagram that introduces a third mode, the "Pragmatic" (Figure 3.1).

The original work on the definition of the "Selective" mode was based upon an analysis of the environmental conditions of the temperate regions of the northern hemisphere, where seasonal

Table 3.1. General characteristics of exclusive and selective mode buildings (Hawkes, 1980)

Exclusive Mode	Selective Mode
Environment is automatically controlled and is predominantly artificial.	*Environment* is controlled by a combination of automatic and manual means and is a variable mixture of natural and artificial.
Shape is compact, seeking to minimize the interface between exterior and interior environments.	*Shape* is dispersed, seeking to maximize the potential collection and use of ambient energy.
Orientation is disregarded.	*Orientation* must be carefully observed.
Windows are generally restricted in size.	*Window* size varies with orientation, large on south-facing facades, restricted to the north.
Energy is primarily from generated sources and is used throughout the year in relatively constant quantities.	*Energy* combines ambient and generated sources. The use varies seasonally, with peak demand in winter and "free-running" operation in summer.

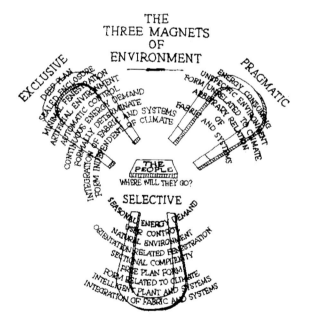

3.1.
The three magnets of environment. (Hawkes, 1996)

differences are marked and where distinctions of orientation are of significance in design. Later work sought to extend and generalize the application of "Selective" principles to the global scale. This led to a comprehensive restatement of the principles in which the significance of accounting for local climate and regional practices is emphasized. (Table 3.2).

Table 3.2. Global characteristics of "Selective" design (Hawkes, McDonald & Steemers, 2002)

Internal environment	Standards are related to the local climate. Emphasis is upon the maximisation of natural light. Primary temperature control is by the building fabric. There is spatial and temporal diversity of conditions. Control is by the occupant.
Built form	Related to the specific climate. Influenced by regional practice. Cross-section a key element of environmental response.
Orientation	Related to the specific climate. Influenced by regional practice. Knowledge of solar geometry is essential.
Fenestration	Related to the specific climate. Influenced by regional practice. Window design must balance the relationship between the thermal and luminous environments.
Energy sources	Energy should be primarily from ambient sources: exploiting natural lighting, useful solar gains and natural ventilation, where appropriate. Mechanical systems for heating, cooling, ventilation and lighting should be regarded as supplementary to the primary control provided by the selective built form. Direct use of renewable energy through the use of water-heating and photovoltaic systems should be considered.

3.3   NATURE AND ARCHITECTURE

The relationship between the external and internal environment lies at the heart of the distinction between the "Exclusive" and "Selective" modes of environmental control. In discussing "the contingencies of climate and the temporally inflected qualities of local light," Frampton (1983) has written,

> *The generic window is obviously the most delicate point at which these two natural forces impinge upon the outer membrane of the building, fenestration having an innate capacity to inscribe architecture with the character of a region and hence to express the place in which the work is situated.*

The key strategy of "Exclusive" design is to minimize the impact of the naturally occurring climate on the internal conditions of a building. The combination of a sealed envelope and mechanical and electrical service systems allows the internal environment to be almost entirely artificial. Nature is held at bay in the interests of notions of "precision" of control and "efficiency" of operation, machines have taken over the environmental function. But, until that moment at some point in the twentieth century at which this separation was first realized, all architecture had been fundamentally concerned to enter into a kind of treaty with the natural environment in order to achieve conditions under which the needs, practical, cultural and symbolic, of humankind could be met. In that sense all of these buildings were environmentally "Selective."

3.2.
The Pantheon, Rome.

The success of this strategy can be amply demonstrated through the built evidence of architectural history. Buildings for all purposes, through their form, construction and detail, become a kind of representation of the conditions within which they were created. The Pantheon at Rome (Figure 3.2) demonstrates an understanding of the relationship between the brightness of the sky and its potential to illuminate a large volume of space.

In Christopher Wren's Library at Trinity College, Cambridge (Figure 3.3), the proportional relationship between the windows, set high in the walls, and the cross section of the space ideally lights both books and readers.

Over two centuries later, Henri Labrouste followed this model in his design for the Biblioteque St. Genevieve in Paris, although, by this date, the technologies of both structural engineering and space heating had developed considerably and found their place in the architecture of the building (Figure 3.4).

3.3.
Library, Trinity College, Cambridge.

3.4.
Biblioteque St. Genevieve, Paris.

The complexity of fenestration on the south façade of Mackintosh's Glasgow School of Art (Figure 3.5) is evidence of the complex relationships that occur between the internal organization and functional complexity of a building such as this and the external environment. It is interesting to note here that, amongst the diverse window openings, there are also a number of louvred grilles that serve the building's ventilation system.

This connection between the inside and outside continued to inform much architecture of significance throughout the twentieth century. In the Scandinavian tradition the works of Erik Gunnar Asplund and Alvar Aalto consistently reveal an acute understanding of nature and its specific manifestation in climate. Asplund's extension to the Gothenburg Law Courts (Figure 3.6) shows this sensibility at work in the heart of the city and Aalto's

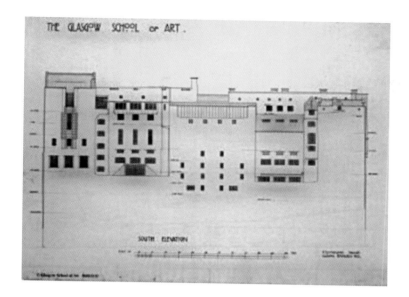

3.5.
Glasgow School of Art.

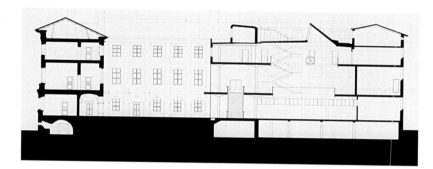

3.6.
Gothenburg Law Court
extension.

Saynatsalo Town Hall (Figure 3.7) is beautifully adapted to its context on a wooded island. In both cases the cross section is a key element in controlling the interaction between inside and outside and, at these northern latitudes, orientation is a controlling factor.

As a final example, consider the works of Louis I. Kahn. The Unitarian Church at Rochester, New York (Figure 3.8) and the Kimbell Art Museum at Fort Worth (Figure 3.9) demonstrate the application of consistent principles in designing buildings that are functionally different and located in quite distinct climates – New York State and Texas, at latitudes 41°N and 33°N, respectively. As in the works of Asplund and Aalto the cross section plays a vital role, but the formal outcomes, one building with an exaggerated silhouette, the other with a consistent and repetitive spatial and structural system, are radically different from each other and from the work of the Scandinavian masters.

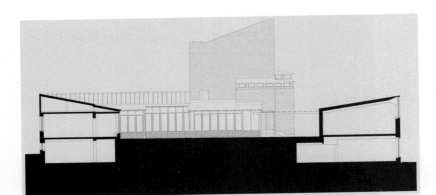

3.7.
Saynatsalo Town Hall.

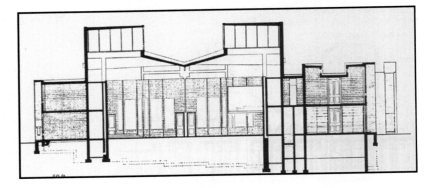

3.8.
Rochester Unitarian Church.

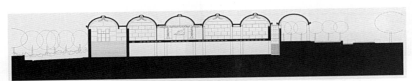

3.9.
Kimbell Art Museum.

All of these examples show that, across centuries and continents, buildings have used form and materiality to connect and transform the relationship between the external and internal climates. The argument here is that this strategy lies at the heart of architecture, as both a technical and cultural phenomenon. When this link is severed by the imposition of a sealed envelope and the introduction of environmental machines something of the essence of architecture is forfeited. But it is vital to stress that "Selective" design is not a reactionary strategy. As needs and techniques have evolved, the potentiality of new circumstances has led new solutions to old problems and the discovery of appropriate solutions to new needs.

Writing in 1953 of Kahn's Yale Art Gallery, Vincent Scully (1953) referred to the sense of "sombre and archaic tension" that he found in the design. This judgement might be applied to all of the works of Kahn's maturity. The Kimbell Art Museum is lit by an aperture at the apex of a vault, as is the Pantheon, but Kimbell is entirely a building of its time adopting the technologies of the twentieth century in meeting the stringent environmental needs of the modern art museum. Of necessity this involves air-conditioning, but this finds its place within a conception of the envelope that is rooted in the historical method of architecture. It is from this that Scully's "sombre and archaic tension" derives.

3.4   COMFORT AND CLIMATE

As life has arisen through the hidden aspects of natural laws, so for better or worse the rules of nature command that life make a close adjustment to natural background. The setting is impartial; it can be cruel or kind, but all living species must either adapt their physiology through selection or mutations, or find other defences against the impact of the environment. (Olgyay, 1963)

The mechanization of environmental control that is represented by the "Exclusive" mode presupposes that human comfort is a matter of narrow, quantified standards of heat, light and sound and that these standards are universal. The "Selective" environment, however, is founded on an alternative approach in which comfort is acknowledged to be a complex phenomenon that involves spatial and temporal variation and both long- and short-term adaptation by the users of buildings.

A key development in research into the theoretical basis of environmental design has been the idea of the "adaptive" model. Michael Humphreys (1997) is one of the leading advocates of this approach and he has shown that people take a whole range of actions to secure satisfactory conditions proposing that, "If change occurs such as to produce discomfort, people react in ways which tend to restore comfort." Within this philosophy Humphreys makes a number of specific proposals for design action. These coincide with the broad principles of "Selective" design. In paraphrase these include:

- The environment should be predictable: the occupants should know what to expect.
- The environment should be "normal": it should be within the range acceptable within the social circumstances in that society and climate.
- Provide thermal variety, where this is appropriate.
- Where people must be at fixed location, provide adequate control of their immediate environment.
- Avoid sudden changes of temperature.

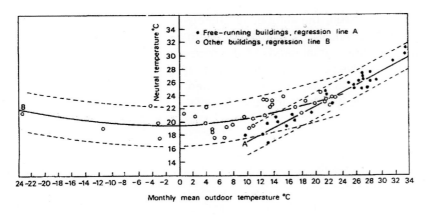

Humphreys has shown that there is a strong correlation between ambient temperatures and acceptable indoor conditions. His work also suggests that people are more tolerant of conditions in what he terms *free-running* buildings – characteristic of "Selective" designs – than in those with the highly controlled environments provided by "Exclusive" design (Figure 3.10).

The significance of Humphreys' work is its explicit demonstration that environmental conditions in buildings should be, within limits, spatially and temporally diverse and that there is clear evidence that expectations are related to regional conditions.

3.5 EXEMPLARS OF "SELECTIVE" DESIGN

A "Selective" building, by definition, enjoys an intimate relationship with its immediate environment. That environment is explicitly and objectively characterized by the physical description of the local climate and by the set of conditions within which comfort may be achieved. However, these parameters are always subject to elaboration and interpretation in the light of the cultural pressures that are crucial to the production of a work of architecture. Whilst many important facts about buildings can be presented in the formulations of building science, these inevitably remain in the realm of the abstract. One of the most potent ways in which architectural knowledge is transmitted, is by reference to specific built cases. Here it is possible to address the complex relationships of the technological and the cultural. The following exemplars, drawn from recent practice, are offered as instances of this.

The Library at Darwin College, Cambridge, by Jeremy Dixon and Edward Jones, was completed in 1994 (Figure 3.11). Any new building in the core of an historic city carries obligations to its setting, so the form and language of the building are specific to its site

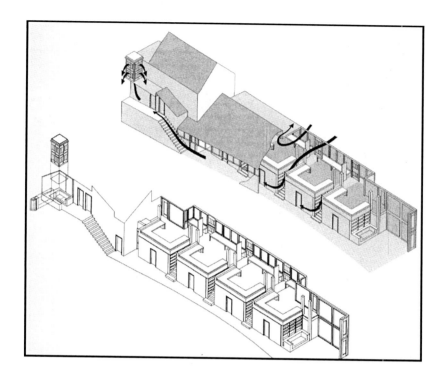

3.11.
Library, Darwin College,
Cambridge.

and programme. Here a narrow site on the bank of the River Cam is occupied by a building that exploits its orientation and form to create a naturally lit and ventilated environment for study. A range of windows facing south over the river admit useful solar heat and combine with a tall vent chimney to the north to provide fresh air and summer cooling.

The Canadian architects Patkau consistently address environmental questions in their work. Strawberry Vale School in Victoria (Figure 3.12) adds to the long line of school designs that occupy a place in the history of environmental architecture. Its elaborate cross section is the product of response to the needs of the educational programme – low spaces for classrooms, high for central, shared resources – and to orientation, with the classrooms facing south into the landscape and service rooms to the north.

One of the most environmentally radical buildings of recent years is the training centre at Mont Cenis, Herne-Sodingen in Germany (Figure 3.13). This adopts an all-enveloping primary structure that provides the first degree of environmental selection, weatherproofing, temperature modification and, through the use of photovoltaic arrays, energy generation. Within this buffer zone, buildings for a wide range of public functions are located, free from the rigours of the external climate, a mini city within an idealised microclimate.

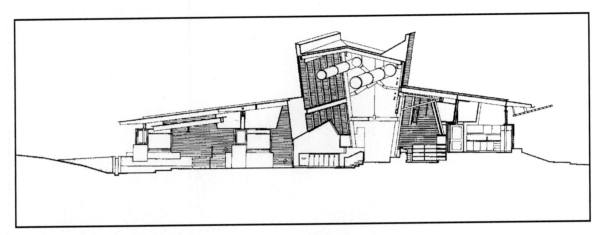

3.12.
Strawberry Vale School, Victoria, British Columbia.

3.13.
Mont Cenis Training Centre,
Herne-Sodingen.

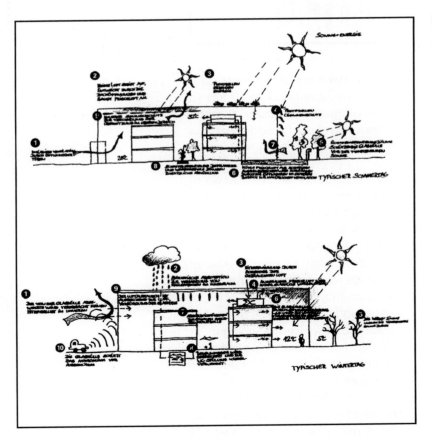

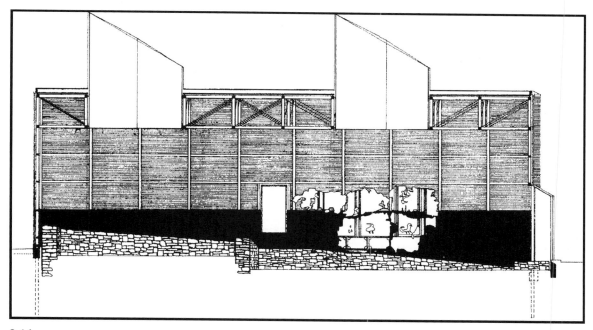

3.14.
Shelter for Roman Remains, Chur.

The work of Peter Zumthor has attracted international attention in recent years. One aspect that has received little attention is his acute sensitivity to environmental questions, in their widest interpretation. The shelter constructed over the remains of a Roman villa at Chur in Switzerland (Figure 3.14) is, in the best sense of the term, a primitive building. It eschews any of the apparatus of mechanical environmental control, with the exception of an artificial lighting installation. The rooflit, timber-louvred enclosure encloses and dramatizes the ancient remains and provides stabilized conditions for their preservation.

Glenn Murcutt's singular practice in Australia has produced a long sequence of environmentally responsive buildings that reinterpret the conditions of Australia through the experience of modernism. As Françoise Fromonot (2003) has observed,

> *Murcutt continues to attach an analysis of natural phenomena to aesthetic implications – to ally the scientific with the picturesque.*

The vast majority of Murcutt's buildings are houses, but, in the few larger projects that he has undertaken, he demonstrates how the principles of domestic design can be equally effectively applied to other uses. The Arthur and Yvonne Boyd Art Centre at Riversdale (Figure 3.15) is the most recent instance of this. Set in a beautiful landscape, the building evinces numerous specific responses to climate and topography. The oversailing roof that is characteristic of

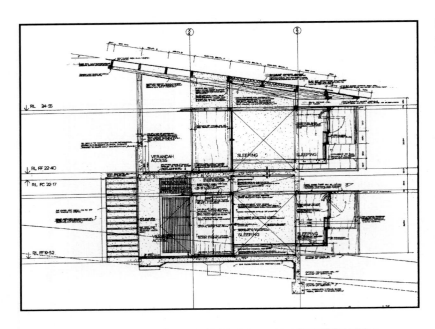

3.15.
Arthur and Yvonne Boyd Art
Centre, Riversdale.

Murcutt's buildings is the primary element of environmental control,
beneath which diverse functions are housed.

The Torrent Pharmaceuticals research laboratory at Ahmedabad
is the product of collaboration between Indian architects Abhikram
and the British architect-environmentalist Brian Ford (Figure 3.16).
The significance of the building is that it applies principles of natural

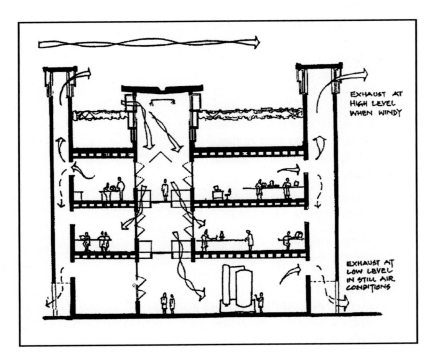

3.16.
Torrent Pharmaceuticals
Laboratory, Ahmedabad.

cooling, ventilation and lighting to a building type that conventionally relies upon mechanical solutions for environmental control. In addition, it achieves this in a hot climate that is inherently hostile. The environmental strategy has a profound influence on the form of the building, which becomes a potent symbol of both its technology and its ideology.

3.6   CONCLUSION

This chapter has sought to show that the production of environmentally responsive – sustainable – architecture should be seen in a continuing tradition in which the form and nature of the building envelope retains its historical function as the primary agent of climate response. The evidence of history shows that the form and material of a building and the size and location of openings, windows and roof apertures, are essential technical instruments of climate response, but carry equal significance in declaring the nature of specific regional, cultural response to the physical environment. The argument for a "Selective" strategy of environmental design admits the role and value of mechanical systems of heating, cooling, ventilation and illumination, but places these in a responsive relationship to building form and material.

This is, of its nature, a general, *generic*, proposition. The "theory" of "Selective" design had its origins in the specific climatic and cultural setting of the northern European climate and, in that respect was "site specific". The further development of the approach, as presented in *The Selective Environment*, and summarized in this chapter, sought to widen its relevance to the global scale. But, while the scope is global, the outcome is specific, *regional*. The application of the principles of "Selective" design leads to buildings that are, in their technical dimensions, attuned to their specific climate and achieve appropriate standards of environmental comfort with economy of means. Viewed from the wide, historical perspective adopted in this chapter the tropical works of Geoffrey Bawa, Ken Yeang, Glenn Murcutt, amongst others, may all be defined as "Selective". In their differing ways these architects resist the common acceptance of the air-conditioner, defined by Frampton as the "antagonist of rooted culture," and thereby produce architecture that celebrates both the climatically responsive and culturally appropriate.

ACKNOWLEDGEMENTS

Much of the work described in this chapter was undertaken in collaboration with Jane McDonald and Koen Steemers at the Martin

Centre at Cambridge as part of a project funded by the Mitsubishi Corporation. Further work has been supported by the award of a Leverhulme Emeritus Research Fellowship, 2002–2004, to study "The Environmental Function of Architecture," the outcome of which is published as *The Environmental Imagination*, by Taylor & Francis, London, 2006.

REFERENCES

Banham, R. (1969) *The Architecture of the Well-tempered Environment*, London: The Architectural Press.

Frampton, K. (1983) Towards a Critical Regionalism: Six Points for an Architecture of Resistance, in Foster, H. (ed.), *Postmodern Culture*, London and Concord, MA: Pluto Press.

Fromonot, F. (2003) *Glenn Murcutt: Buildings and Projects, 1962–2003*, London and New York: Thames and Hudson.

Hawkes, D. (1980) Building Shape and Energy Use, in Hawkes, D. and Owers, J. (eds), *The Architecture of Energy*, London: Longmans.

Hawkes, D. (1996) *The Environmental Tradition: Studies in the Architecture of Environment*, London: E. & F.N. Spon.

Hawkes, D., McDonald, J. and Steemers, K. (2002) *The Selective Environment: an Approach to Environmentally Responsive Architecture*, London: Spon Press.

Humphreys, M. (1997) An Adaptive Approach to Thermal Comfort Criteria, in Clements-Croome, D. (ed.), *Naturally Ventilated Buildings: Buildings for the Senses, the Economy and Society*, E. & F.N. Spon.

Le Corbusier (1930) *Precisions on the Present State of Architecture and City Planning*, Crès et Cie., Paris. English translation, Boston, MA: MIT Press, 1991.

Olgyay, V. (1963) *'Design with Climate: Bioclimatic Approach to Architectural Regionalism'*, Princeton, NJ: Princeton University Press.

Scully, V. (1953) Sombre and Archaic: Expressive Tension, *Yale Daily News*, 6 November.

Tzonis, A., Lefaivre, L. and Stagno, B. (2001) *Tropical Architecture: Critical Regionalism in the Age of Globalization*, Chichester: Wiley-Academy.

4 GREEN DESIGN IN THE HOT HUMID TROPICAL ZONE

Ken Yeang

T.R. Hamzah and Yeang Sdn. Bhd.

Abstract

Already well-argued elsewhere is the case for architects and engineers to design our man-made environment with sustainable and "green" (ecologically respon-sive) design objectives. Simply stated here, the approach is to build with minimal impact on the natural environment, to integrate the built-environment and its systems with the ecological systems (ecosystems) of the locality and if possible, to positively contribute to the ecological and energy productivity of the location. For many designers today, these objectives are regarded as pre-requisites for all their design endeavours.

Keywords

Ecological design, bioclimatic, building configuration, natural ventilation, passive low-energy, sunpath, roofscape, vertical landscaping, materials, waste, ecosystems.

The proposition here, of the intensive building as an ecologically responsive building might well be regarded by some as a conun-drum. Afterall, an intensive building such as the skyscraper is one of the city's most intensive building type (besides others like the shopping mall or the convention center). Such buildings, in compar-ison with other smaller experimental low energy ecological buildings elsewhere, are by all means and purposes not low energy nor self evidently ecological building types by virtue of their enormous size and high consumption of energy and materials.

4.1.
Mewah Oils HQ & Refinery,
Port Klang, Malaysia.

The point then is that regardless of whether one is for or against intensive urban buildings, in reality these building types will not simply go away. For example, current trends in fact, indicate that the skyscraper as a building type shall continue to be built into the next millennium as well and to be in existence in most of the world's cities unless man invents another economical alternative, or until radical changes are effected in planning administration or to the current high trend of rural-to-urban migration into cities.

The pressing key issue is how can the designer today design these massive building types to be ecologically responsive? This issue has to be urgently addressed. The alternative would be a situation where the rapidly developing cities continue to be filled with multitudes of high energy-consuming, waste-producing, polluting and intensive buildings. Our efforts must be urgently directed to designing ecologically responsive skyscrapers and other intensive building types to reduce their aggregate negative impacts on the biosphere.

In this chapter, the key factors for the green design of the intensive building are explained in the form of a simple General Systems model of the crucial interactions that any built system will have on the natural environment. As a model for design, it provides a check of those items to be considered if design is to be regarded as ecological. Illustrated are a number of built and designed precedents, using case studies to explain the design principles and ideas.

Specifically, this chapter discusses the ecological benefits of the following: passive low energy strategies and related bioclimatic design principles as a subset of ecological design (such as principles on the location of the service cores, natural ventilation of spaces, orientation of builtform, building configuration options, sunshading, wind-scoops, atriums, skycourts, etc.), the use of transitional spaces, façade design, roofscape design, building M&E Systems design, wind and natural ventilation, natural light and lighting systems, vertical landscaping, energy embodiment in buildings (primary and delivered), materials strategy (reduce, reuse, recycling, reintegration, etc.), urban design implications, life cycle impacts, etc.

1. Saving our environment is the most vital issue that humankind must address today, feeding into our fears that this millennium may be our last.

   For the designer, the compelling question is: how do we design for a sustainable future? This question, similarly, concerns industry. Companies now anxiously seek to understand the environmental consequences of their business, envision what their business might be if it were sustainable, and seek ways to realize this vision with ecologically benign strategies, new business models, production systems, materials, and processes.

If we have an ecologically responsive built environment, it will likely change the way we work and our present ecologically profligate way of life.

What is green design and particularly in the tropical zone where the climate is characterized by high humidity, generally high temperatures throughout the year and high rainfall? Presented here are some propositions addressing this idea.

2. The ecological approach to our businesses and design is ultimately about environmental integration.

If we integrate our business processes and design and everything we do or make in our built environment (which by definition consists of our buildings, facilities, infrastructure, products, refrigerators, toys, etc.) with the natural environment in a seamless and benign way, there will be no environmental problem whatsoever.

Simply stated, ecodesign is designing for bio-integration. This can be regarded at three aspects: physically, systemically and temporally. Successfully, achieving aspects is, of course, easier said than done, but herein lies our challenge.

3. We start by looking at nature. Nature without humans exists in stasis. Can our businesses and our built environment imitate nature's processes, structure, and functions, particularly of its ecosystems? A key characteristic of the ecology of the hot humid tropics is the indigenous tropical rain forest vegetation much of which is in advanced state of succession with a high biodiversity.

In such ecosystems, where there is a high level of complexity and trophic levels, biointegration is particularly important. For instance, ecosystems have no waste. Everything is recycled within. Thus by imitating this, out built environment will produce no wastes. All emissions and products are continuously reused, recycled within, and eventually reintegrated with the natural environment, in tandem with efficient uses of energy and material resources. Designing to imitate ecosystems is ecomimesis. This is the fundamental premise for ecodesign. Our built environment must imitate ecosystems in all respects.

4. Nature regards humans as one of its many species. What differentiates humans is their capability to force large scale devastative changes to the environment. Such changes are often the consequences of manufacturing, construction, and other activities (e.g. recreation and transportation).

5. Our built forms are essentially enclosures erected to protect us from the inclement external weather, enabling some activity (whether residential, office, manufacturing, warehousing, etc.) to take place. This is particularly crucial in the tropical climatic zone.

    Ecologically, a building is just a high concentration of materials on a location (often using nonrenewable energy resources) extracted and manufactured from some place distant in the biosphere, transported to that location and fabricated into a built form or an infrastructure (e.g. roads and drains), whose subsequent operations bear further environmental consequences and whose eventual after-life must be accommodated.

6. There is also much misperception about what is ecological design today. We must not be misled by the popular perception that if we assemble in one single building enough eco-gadgetry such as solar collectors, photo-voltaics, biological recycling systems, building automation systems and double-skin façades, we will instantaneously have an ecological architecture.

    The other misperception is that if our building gets a high notch in a green-rating system, then all is well. Of course, nothing could be further from the truth. Worse, a self-complacency sets in whereupon nothing further is done to improve environmental degradation.

    Although these eco-gadgetry and technological systems are relevant experiments, perhaps, towards an eventual ecologically responsive built environment, their assembly into one single building does not make it automatically ecological.

7. In a nutshell, ecodesign is designing the built environment as a system within the natural environment. The system's existence has ecological consequences and its sets of interactions, being its inputs and outputs as well as all its other aspects (such as transportation, etc.) over its entire life cycle, must be benignly integrated with the natural environment.

8. Ecosystems in a biosphere are definable units containing both biotic and abiotic constituents acting together as a whole. From this concept, our businesses and built environment should be designed analogously to the ecosystem's physical content, composition and processes. For instance, besides regarding our architecture as just art objects or as serviced enclosures, we should regard it as artifacts that need to be operationally and eventually integrated with nature (see (4) above).

9. As is self-evident, the material composition of our built environment is almost entirely inorganic, whereas ecosystems contain a complement of both biotic and abiotic constituents, or of inorganic and organic components.

Our myriad of construction, manufacturing and other activities are, in effect, making the biosphere more and more inorganic, artificial and increasingly biologically simplified. To continue without balancing the biotic content means simply adding to the biosphere's artificiality, thereby making it increasingly more and more inorganic. Exacerbating this are other environmentally destructive acts such as deforestation and pollution. This results in the biological simplification of the biosphere and the reduction of its complexity and diversity.

We must first reverse this trend and start by balancing our built environment with greater levels of biomass, ameliorating biodiversity and ecological connectivity in the built forms and complementing their inorganic content with appropriate biomass.

10. We should improve the ecological linkages between our designs and our business processes with the surrounding landscape, both horizontally and vertically. Achieving these linkages ensures a wider level of species connectivity, interaction, mobility and sharing of resources across boundaries. Such real improvements in connectivity enhance biodiversity and further increase habitat resilience and species survival.

    Providing ecological corridors and linkages in regional planning is crucial in making urban patterns biologically more viable.

    Besides improved horizontal connectivity, vertical connectivity within the built form is also necessary since most buildings are not single- but multi-story. Design must extend ecological linkages upwards within the built form to its roofscapes.

11. More than enhancing ecological linkages, we must biologically integrate the inorganic aspects and processes of our built environment with the landscape so that they mutually become ecosystemic (see (4) above). We must create "human-made ecosystems" comptatible with the ecosystems in nature.

    By doing so, we enhance human-made ecosystems' abilities to sustain life in the biosphere.

12. Ecodesign is also about discernment of the ecology of the site. Any activity from our design or our business takes place with the objective to physically integrate benignly with the ecosystems (see (4) above).

    Particularly in site planning, we must first understand the properties of the locality's ecosystem before imposing any intended human activity upon it. Every site has an ecology with a limiting capacity to withstand stresses imposed upon it, which if stressed beyond this capacity, becomes irrevocably damaged. Consequences can range from minimal localized impact (such as the clearing of a small land area for access), to the total devastation of the entire land area (such as the clearing of all trees and vegetation, leveling the topography, diversion of existing waterways, etc.).

13. To identify all aspects of this carrying capacity, we need to carry out an analysis of the site's ecology.

    We must ascertain its ecosystem's structure and energy flow, its species diversity and other ecological properties. Then we must identify which parts of the site (if any) have different types of structures and activities, and which parts are particularly sensitive. Finally, we must consider the likely impacts of the intended construction and use.

14. This is, of course, a major undertaking. It needs to be done over the entire year and in some instances over several years. To reduce this lengthy effort, landscape architects developed the "layer-cake" method, or a sieve-mapping technique of landscape mapping. This enables the designer to map the landscape as a series of layers in a simplified way to study its ecology.

    As we map the layers, we overlay them, assign points, evaluate the interactions in relation to our proposed land use and patterns of use, and produce the composite map or guide our planning (e.g. the disposition of the access roads, water management, drainage patterns and shaping of the built form(s), etc.).

    We must be aware that the sieve-mapping method generally treats the site's ecosystem statically and may ignore the dynamic forces taking place between the layers and within an ecosystem. Between each of these layers are complex interactions. Thus, analyzing an ecosystem requires more than mapping. We must examine the inter-layer relationships.

15. We must also look into ways to configure the built forms and operational systems for our built environment and our businesses as low-energy systems.

    In addressing this, we need to look into ways to improve internal comfort conditions. There are essentially five modes: Passive Mode (or bioclimatic design), Mixed Mode, Full Mode, Productive Mode and Composite Mode, the latter being a composite of all the proceeding.

    Designing means looking at Passive Mode strategies first, then Mixed Mode to Full Mode, Productive Mode and to Composite Mode, all the while adopting progressive strategies to improve comfort conditions relative to external conditions.

    Meeting contemporary expectations for comfort conditions, especially in manufacturing, cannot be achieved by Passive Mode or by Mixed Mode alone. The internal environment often needs to be supplemented by using external sources of energy, as in Full Mode.

Full Mode uses electro-mechanical systems or M&E (mechanical and electrical) systems to improve the internal conditions of comfort, often using external energy sources (whether from fossil fuel derived sources or local ambient sources).

Ecodesign of our buildings and businesses must minimize the use of nonrenewable sources of energy. In this regard, low-energy design is an important design objective but not the only design objective in ecological design.

16. Passive Mode is designing for improved comfort conditions over external conditions without the use of any electro-mechanical systems. Examples of Passive Mode strategies include adopting appropriate building configurations and orientation in relation to the locality's climate, (and in this instance, the tropical climatic conditions), appropriate façade design (e.g. solid-to-glazed area ratio and suitable thermal insulation levels, use of natural ventilation, use of vegetation, etc.).

    The design strategy for the built form must start with Passive Mode or bioclimatic design. This can significantly influence the configuration of the built form and its enclosural form. Therefore, this must be the first level of design consideration in the process, following which we can adopt other modes to further enhance the energy efficiency.

    Passive Mode requires an understanding of the climatic conditions of the locality, then designing not just to synchronize the built form's design with the local meteorological conditions, but to optimize the ambient energy of the locality into a building design with improved internal comfort conditions without the use of any electro-mechanical systems. Otherwise, if we adopt a particular approach without previously optimizing the Passive Mode options in the built form, we may well have made non-energy-efficient design decisions that will have to correct with supplementary Full Mode systems. This would make nonsense of designing for low energy.

    Furthermore if the design optimizes its Passive Modes, it remains at an improved level of comfort during any electrical power failure. If we have not optimized our Passive Modes in the built form, then when there is no electricity or external energy source, the building may be intolerable to occupy.

    Responding by design to the climate conditions, sun-path, wind-rose is particularly important to this tropical zone.

17. Mixed Mode is where we use some electro-mechanical (M&E) systems. Examples include ceiling fans, double façades, flue atriums and evaporative cooling.

18. Full Mode is the full use of electro-mechanical systems, as in any conventional building. If our users insist on having consistent comfort conditions throughout the year, the designed system heads towards a Full Mode design.

    It must be clear now that low-energy design is essentially a user driven condition and a life-style issue. We must appreciate that Passive Mode and Mixed Mode design can never compete with the comfort levels of the high-energy, Full Mode conditions.

19. Productive Mode is where the built system generates its own energy (e.g. solar energy using photo-voltaics, or wind energy).

    Ecosystems use solar energy, which is transformed into chemical energy by the photo-synthesis of green plants and drives the ecological cycle. If ecodesign is to be ecomimetic, we should seek to do the same. At the moment the use of solar energy is limited to various solar collector devices and photovoltaic systems.

    In the case of Productive Modes (e.g. solar collectors, photo-voltaics and wind energy), these systems require sophisticated technological systems. They subsequently increase the inorganic content of the built form, its embodied energy content and its use of material resources, with increased attendant impacts on the environment.

20. Composite Mode is a composite of all the above modes and is a system that varies over the seasons of the year. In the hot humid tropical belt, there is less differentiation as there are usually no mid-seasons.

21. Ecodesign also requires the designer to use green materials and assemblies of materials, and components that facilitate reuse, recycling and reintegration for temporal integration with the ecological systems (see (4) above).

    We need to be ecomimetic in our use of materials in the built environment. In ecosystems, all living organisms feed on continual flows of matter and energy from their environment to stay alive, and all living organisms continually produce wastes. Here, an ecosystem generates no waste, one species' waste being another species' food. Thus matter cycles continually through the web of life. It is this closing of the loop in reuse and recycling that human-made environment must imitate (see (2) above).

    We should unceremoniously regard everything produced by humans as eventual garbage or waste material. The question for design, businesses and manufacturing is: what do we do with the waste material?

    If these are readily biodegradable and biointegration is more rapid in the hot humid tropical belt, they can return into the environment through decomposition, whereas the other generally inert wastes need to be deposited somewhere, currently as landfill or pollutants.

Ecomimetically, we need to think about how a building, its components and its outputs can be reused and recycled at the outset in design before production. This determines the processes, the materials selected and the way in which these are connected to each other and used in the built form.

For instance, to facilitate reuse, the connection between components in the built form and in manufactured products needs to be mechanically joined for ease of demountability. The connection should be modular to facilitate reuse in an acceptable condition.

22. Another major design issue is the systemic integration of our built forms and its operational systems and internal processes with the ecosystems in nature.

This integration is crucial because if our built systems and processes do not integrate with the natural systems in nature, then they will remain disparate, artificial items and potential pollutants. Their eventual integration after their manufacture and use is only through biodegradation. Often, this requires a long term natural process of decomposition.

While manufacturing and designing for recycling and reuse within the human-made environment relieves the problem of deposition of waste, we should integrate not just the inorganic waste (e.g. sewage, rainwater runoff, waste water, food wastes, etc.) but also the inorganic ones as well.

23. We might draw an analogy between ecodesign and prosthetics in surgery.

Ecodesign is an essential design that integrates our artificial systems both mechanically and organically, with its host system being the ecosystems. Similarly, a medical prosthetic device has to integrate with its organic host being – the human body. Failure to integrate well will result in dislocation in both.

By analogy, this is what ecodesign in our built environment and in our businesses should achieve: a total physical, systemic and temporal integration (see (4) above) of our human-made, built environment with our organic host in a benign and positive way.

24. Discussion here on some of the key issues will help us approach the ecological design of artifacts and our businesses to be environmentally responsive.

There are, of course other aspects. There are still a large number of theoretical and technical problems to be solved before we have a truly ecological built environment whether in the tropical, temperate or cold climatic zones.

BIBLIOGRAPHY

Hawkes, D. and Wayne, F. (2002) *Architecture, Engineering and Environment.* London: Laurence King Pub. in association with Arup.

Olgyay, V.W. and Soontorn, B. (2000) The Shinawatra University: Design for the millennium, in Steemers, K. and Yannas, S. (eds), *Architecture, City, Environment: Proceedings of PLEA International Conference 2000, Cambridge, UK,* London: James & James.

Steemers, K. and Yannas, S. (eds) (2000) *Architecture, City, Environment: Proceedings of PLEA International Conference 2000, Cambridge, UK,* London: James & James.

Tzonis, A., Lefaivre, L. and Stagno, B. (eds) (2001) *Tropical Architecture: Critical Regionalism in the Age of Globalization,* Chichester: Wiley-Academy, with Fonds, Prince Claus Fund for Culture and Development, The Netherlands.

University of Chicago Press (1993) *The Chicago Manual of Style,* 14th ed. Chicago: University of Chicago Press.

Yeang, K. (1996) The *Skyscraper, Bio-Climatically Considered; A Design Primer.* London: Academy Editions.

Part II

HIGH-RISE HIGH-DENSITY LIVING

5 SOCIO-ENVIRONMENTAL DIMENSIONS IN TROPICAL SEMI-OPEN SPACES OF HIGH-RISE HOUSING IN SINGAPORE

Joo-Hwa Bay, Na Wang, Qian Liang and Ping Kong

Department of Architecture, National University of Singapore

Abstract

What are the relationships of community, tropical environment and semi-open spaces in the high-rise high-density housing? The social and environmental aspects of the veranda spaces of the traditional kampong (village) houses in the tropical regions are integrated and sustainable. Is it possible to have community living in semi-open spaces such as entrance forecourts and corridor spaces in high-rise apartments? What are the environmental conditions that permit this? What are the relationships of socio-climatic aspects with plants in sky-gardens and sizes of semi-open spaces? All these are discussed in the case of Bedok Court condominium, with comparison with a typical public housing block in Singapore.

Keywords

Sustainability, environment, design guidelines, high-rise high-density, housing, community, semi-open space, socio-climatic, tropical.

5.1 INTRODUCTION

Traditionally, the *angung* and *serambi* (both veranda spaces) of the kampong (village) houses in the region (Chen, 1998; Lim, 1981) are environmentally conducive for various social activities, including children's play, dining and receiving guests (Figure 5.1). Besides providing shade for the interior, reducing the direct heat from the sun, the veranda spaces are highly visible from the kampong streets, encouraging high levels of familiarity and neighbourliness

5.1.
Indigenous traditional house in Malaysia with the anjung porch and the serambi behind it. (*Source:* J.H. Bay)

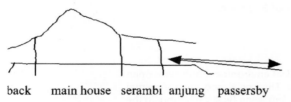

back   main house serambi anjung   passersby

5.2.
Highly visible semi-open veranda of traditional Kampong house in the region. (*Source:* J.H. Bay)

(Figure 5.2). A similar socio-climatic phenomenon is also observed in Bedok Court condominium that has a forecourt (veranda) to each apartment in the high-rise high-density context, where almost a hundred percent of the residents expressed that they have a strong sense of community and security (Figure 5.3).

What is the implication of this socio-environmental phenomenon of the veranda in relation to the issue of sustainability of community and environment in the high-rise high-density context of Singapore?

5.1.1 Quantity and quality of high-rise living in Singapore

Singapore's public high-rise housing (Figure 5.4) developed rapidly since independence in 1965 to accommodate an astonishing 84% of the population today (HDB, 2004). The remaining 11% live in private high-rise condominiums (URA, 2004). Only 5%, mainly from the more affluent strata of Singapore society, live in so-called "landed properties" such as a detached, a semi-detached or a terrace house, and enjoy the use of a private garden.

Some of the pressing questions are: how can community life be encouraged; how can natural ventilation be improved and the

5.3.
View of typical semi-open forecourt from common corridor, Block 1, Bedok Court. (*Source:* J.H. Bay)

5.4.
Typical public housing partment blocks in Singapore. (*Source:* J.H. Bay)

reliance on air-conditioning be reduced; how can the heat island effect be reduced; how can energy consumption, CO_2 emissions be cut and other air pollution be reduced; how can vegetation be increased and how can some of the benefits of "landed living" be made available to apartment dwellers?

Few architectural critics have attempted to analyze the environmental and social characteristics of life in the island's high-rise apartments together. Most books and journal articles[1] focus exclusively on the garden houses of the wealthy. While many would prefer to stay on the ground, land scarcity, as in many cities, results in many efficient typologies of high-rise housing that maximize and manage quantity of people in highly built-up environments.

The issue now is how to provide sustainable social and environmental qualities.

5.1.2 Social and environmental aspects studied separately

Bay (2005) noted that the social aspect has often been studied independently from the environmental aspect. The tropical veranda has been studied as a device for providing shade, reducing cooling load and improving thermal comfort (for example, Yeang, 1996; Hyde, 2000), while others have discussed the feasibility of developing architectural language based on a "bioclimatic" and related environmental design approach (for example, Olgyay and Olgyay, 1963; Yeang, 1996; Hagan, 2001). Lim (1981) compared the traditional Malay house and the modern housing-estate house, observing the differences of lifestyle and the lack of casual semi-open spaces, but did not discuss how aspects of the kampong might be adapted for high-rise high-density living. Gehl (1996) discussed "life between buildings" as street life on the ground, but stopped short of considering the possibility of creating "street life" at upper levels in high-rise apartment developments. There has been little discussion, on the relationship between environmental design and the development of community in high-rise housing in the Tropics.

This chapter first summarizes the overall socio-climatic relationships and success found in the forecourts of Bedok Court condominium, discussed in a previous paper (Bay, 2005). Then it advances the discussion with newer detailed findings (including part of a comparative study of this case with a public housing block), concerning:

(a) the impact of shading and ventilation in achieving thermal comfort in the semi-open forecourt (the effect on the interior of the apartment is not discussed here);
(b) the relationship of gardening, social and environmental benefits;
(c) the optimal size of a typical forecourt for a successful facilitation of the social and environmental benefits.

5.2 COMMUNITY AND A PIECE OF GREEN IN THE SKY

Bedok Court was designed by Cheng Jian Fenn of Design Link Architects, Singapore, in 1982 and completed in 1985. It comprises 280 apartments, distributed in three blocks, which vary from 4- to 20-storey heights (Figures 5.5 and 5.6). The area of the site

5.5.
Bedok Court Condominium with large balconies looking and not visible from common corridors. (*Source:* J.H. Bay)

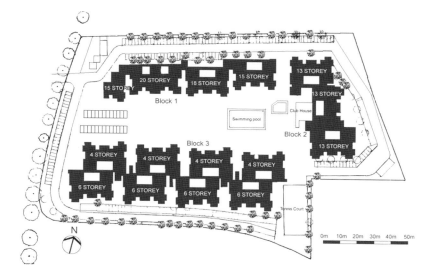

5.6.
Site Plan, Bedok Court. (*Source:* J.H. Bay)

is about 3.4 ha and the total area of the development, including corridors, forecourts and balconies, is about 65 500 m² gross. The resultant density is 300 persons per hectare (or 82 dwellings per hectare) and the floor area ratio is 1.9. The development includes surface car parking lots, landscaped-gardens, tennis courts and a large swimming pool.

What distinguishes Bedok Court is the generous provision of semi-open forecourts and balconies. Typical apartments range in gross area from about 110–220 m², including balconies and fore-courts and excluding common circulation spaces. The larger three bedroom apartments have an internal area of about 110 m² each, while the two- and one-bedroom apartments have about 85 and 55 m² each, respectively, typical of housing development at that

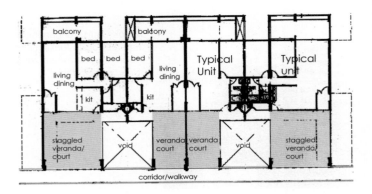

5.7.
Typical Plan, Block 1, Bedok
Court. (*Source:* J.H. Bay)

time. About 30–40% of each apartment is dedicated semi-open space (Figure 5.7).

Interestingly, Cheng chose to develop three different sky-street situations in his scheme. Block 1 (a 15- to 20-storey block) has a single loaded sky-street along its north elevation (Figures 5.7–5.9); Block 2 (a 13-storey block) has a central wing with two side wings placed at right angles to it; and Block 3 (a 4- to 6-storey block) has a double loaded street with a staggered configuration of apartments. In terms of ventilation, Block 1 is the most successful and Block 3 is the least successful. In terms of solar gain limitation, Block 3 is the most successful. In terms of social interaction all blocks are equally successful.

5.2.1 Design Intentions

In an interview by Bay, Cheng stated that his inspiration came from his first-hand experience of the kampongs in Singapore.[2] He wanted to re-create the relaxed friendly atmosphere and strong sense of community and security found in this setting. He suggested that the high degree of visual connectivity of residents in their entrance porch spaces contributed to high levels of social interaction and familiarity, resulting in a strong sense of identity and security. Any visitor would be identified immediately as a stranger.

The architect also wanted to design high-rise apartments, where each resident could own a "piece of green" and a "house" in the sky. In Singapore terraced, semi-detached and single houses are called "landed properties", and apartments are called "non-landed properties". For Bedok Court Cheng wanted to give each dwelling unit an entrance garden similar to that of a "landed dwelling". He mentioned that in the land-scarce Singapore, it would be important to create the possibility for people to own a new kind of "landed

TYPICAL   SECTION   B-B - BLOCK 1
SCALE 1:100

5.9.
Multi-level visibility and comfort in forecourts, Block 1, Bedok Court. (*Source:* J.H. Bay)

5.10.
Semi-open space of Bedok
Court, a piece of green in the
sky. (*Source:* Authors)

property in the sky", where buying an apartment would be likened to buying a "landed dwelling" (Figure 5.10).

Cheng also mentioned that he had been very much influenced by the writings of Jacobs (1962) and her assertion that the modern city needs a vital street life. However, he went on to admit that, as in the original kampongs, some degree of privacy would have to be sacrificed in order to gain the necessary familiarity and trust.

5.2.2   Forecourt, environment and community

The surveys and measurements by Bay[3] to assess the quality of community life and the environment of the forecourt in Bedok Court produced several surprising results:

1. *Preference of the forecourt over other spaces*   A high percentage of interviewed residents (86%) nominated the forecourt/veranda as the most desirable space compared to the interior of the apartment, the balcony, the lift, the lobby, the playground, the swimming pool, and the car parking areas;
2. *High frequency of social activities in forecourts*   Most residents (86%) used this space for social activities, receiving guests, gardening, hobbies, children's play, study group activities and parties, more than once per week;
3. *Thermal comfort condition*   Most residents (80%) felt slightly warm, comfortable, or slightly cool for the three periods of the day, morning, just after noon and evening before dark, for the warmest month of the year;
4. *Daylighting levels*   The majority of the respondents (80%) found the daylighting slightly too bright, comfortable, to slightly dim, and therefore reasonably acceptable;

5. *Acoustic levels*   Most of the respondents reported that they felt comfortable with the overall acoustic levels at the forecourts (most recorded levels were below 65 dBA). Noise was not a problem in the case of Bedok Court. This could be due, in part, to a high level of neighbourliness and tolerance for a certain level of ambient background noise;

6. *Privacy issue*   Majority of residents reported not feeling a lack of privacy (90%). Although there will always be some people who demand total privacy, there seemed to be a large number of people who are willing to trade off reduced "privacy" for increased social contact. In any case, they could still enjoy the usual privacy of the interior of their apartment units, just as in any other apartments;

7. *Sense of belonging, ownership and security*   Almost all the residents interviewed felt a strong sense of belonging, ownership and security.

In interviews Bedok Court residents used the term "kampong" to describe both the spatial system and the community in which they live, and they used the terms "court" and "garden" to refer to the forecourt. The developer adopted even the name, "Bedok Court", because it expressed the unique "court" feature of the design.

Similar to the traditional Malay kampong, each entrance veranda of Bedok Court provided a comfortable environment for social activities, visible to the neighbours. There is a spatial continuity between the semi-privacy of each forecourt and the openness of the common sky-street[4] at various levels, thus functioning as 3-dimensional multi-layered streets within a high-density context, where 66% of the residents knew neighbours on higher or lower levels. The high level of visibility of the daily activities and casual encounters in the corridors and in the forecourts is the key to the intensification of familiarity, which promotes the sense of community.[5] This familiarity was not at the superficial level, but involved home visits on a regular basis.

The shade provided by each veranda together with the vegetation, the wind, and the ventilation in each forecourt provided substantial cooling of the environment to afford the thermal comfort condition. There was no lighting or acoustic problem. The qualities of the environment encourage more social activities, leading to better sense of community, and in turn leading to higher preference for the forecourt environment, strongly inter-connected in a sustainable cycle (Bay, 2004a, 2005).

Introduced in 1986,[6] a year after the completion of Bedok Court, changes in planning and building regulations, effectively discouraged developers from providing semi-open balconies and forecourts.

As a result, most apartments built since 1986 have been conceived as sealed air-conditioned envelopes with almost no outside veranda and balcony space. Singaporeans have retreated further into artificially cooled environments, moving in air-conditioned cars between their air-conditioned homes and air-conditioned offices, shopping in air-conditioned malls and pausing to exercise in air-conditioned gyms.[7] But this life pattern is not only energy expensive – it is also socially isolating. The Bedok Court example suggests that the provision of semi open (non-air-conditioned) spaces can positively encourage a more natural and sustainable way of life and environment, encouraging neighbourliness in high-rise apartments.

5.2.3   Framework for research on semi-open forecourts

Does the above study suggest further lines of investigation in relation to social and environmental dimensions?

Baruch Givoni has observed departures from ASHRAE comfort range in Colima, Mexico (Givoni, 1998, 24–25). For tropical high-rise dwellings, Nyuk-Hien Wong (Wong et al., 2002) has studied the thermal comfort range for the naturally ventilated unit interiors, but there is not yet a similar study for the tropical semi-open spaces. What are the key factors that differ? Boon-Lay Ong (2002) has developed a "green plot-ratio" concept for regulating the amount of planting needed for an urban context to enjoy the environmental, aesthetic, and recreational benefits. This application relates better to top-down decisions of planting, and requires much centralized maintenance. Individual owners manage their own plants in the Bedok Court type of forecourts, and therefore it is worth exploring how these may be increased as a more self-sustaining approach. What can be learnt about the relationship between plants, people, climate and semi-open space that can give clues to increase the desirability of forecourts and gardening?

What are the optimal sizes of semi-open forecourts to achieve a reasonable quality of environment as well as facilitate visible social activities and casual encounters, and thus provide the chance for better community building?

5.3   WIND AND SOLAR RADIATION

Even though the structure of a Bedok Court forecourt (semi-open or semi-enclosed space) comprise a ceiling and a floor plate like the interiors (enclosed spaces) of each apartment, the openings on the sides are substantially much larger then those for the interior rooms. The average radiant temperature measured for the forecourts was generally much lower than that measured for the external environment outside the apartment block and slightly higher than the interior of the apartments.

Liang (2005) argues that wind and solar radiation are the critical factors that affect the thermal comfort, compared to the interior spaces. Therefore the prediction model for thermal comfort of the interior of high-rise tropical housing in Singapore (Wong et al., 2002) cannot be used effectively for the predicting comfort votes for the semi-open forecourts. He surveyed and measured the semi-open forecourts, and adjacent corridors of apartment units of Bedok Court Block 1 (Figures 5.3, 5.7 and 5.8) to study the relative effects of wind and solar radiation on the thermal comfort votes of the residents.

5.3.1 Combined effects of wind and solar radiation

Liang's study shows that there are dramatic differences of the average DBT and average Relative Humidity of the apartment interiors compared to the averages in the forecourts (veranda) and the corridor spaces throughout the day (Figures 5.11 and 5.12). The intensity of solar radiation rose dramatically higher for the narrow corridors at around noon compared to the large and well-shaded forecourts (courtyard) (Figure 5.13). The wind speeds are much stronger at the corridors throughout the day (Figure 5.14).

Even though the wind speed in the forecourts (verandas) are lower, the combined effect with the lower solar radiation due to substantial shading results in many of the comfort votes from slightly warm (+1 to −2 in Figure 5.15). In the tropics it is better to feel slightly cool on the warm side. In the corridors, where the solar

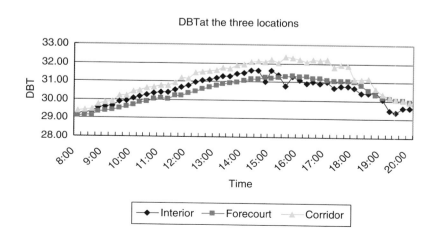

DBT at the three locations

5.11.
Comparison of DBT at the three locations, June 2004, Block 1, Bedok Court. (*Source:* Q. Liang)

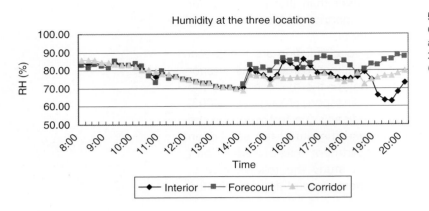

5.12.
Comparison of Relative Humidity at the three locations, June 2004, Block 1, Bedok Court. (*Source:* Q. Liang)

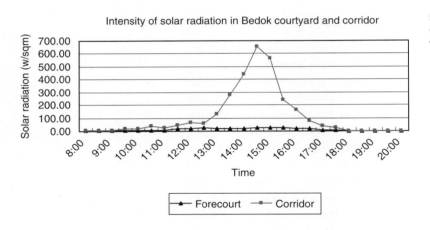

5.13.
Comparison of solar radiation in veranda and corridor, Block 1, Bedok Court. (*Source:* Q. Liang)

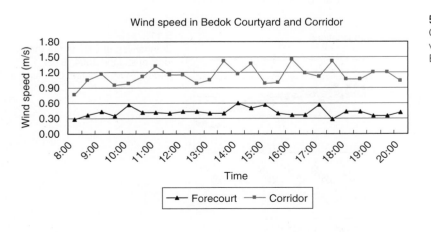

5.14.
Comparison of wind speed in verandas and corridors, Block 1, Bedok Court. (*Source:* Q. Liang)

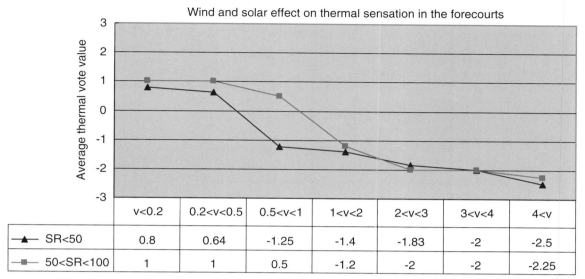

5.15.
Effect of wind and solar effect in the verandas, June 2004, Block 1, Bedok Court. (*Source:* Q. Liang)

radiation is higher, the average votes for comfort rose to very warm and uncomfortably warm if the wind speed drops below 1 m/s.

5.3.2 Solar radiation has stronger effect on thermal comfort votes

The warming effect of solar radiation is therefore more influential in the semi-open space than the cooling effect of wind on thermal comfort. From Liang's study, designers should try to keep solar radiation below 700 W/m^2 for narrow corridor-forecourts to apartment units, where even a low wind speed 0.5 to 1 m/s could still afford thermal comfort vote of slightly warm (Figure 5.16), which would be still acceptable. Measurements of radiation levels of the corridor however show that additional shading devices are needed to keep the radiation below this level for most of the day. For the larger semi-open spaces like the Bedok Court forecourts, measurements of lower than 100 W/m^2 of solar radiation could be achieved with reasonable thermal comfort sensation with minimal wind speed of 0.3–0.6 m/s, and tending towards the cooler side with stronger wind speeds, generally available for the high-rise apartments.

The study shows that the shading of solar radiation provided by the large forecourt is effective and instrumental in the provision of thermal comfort environment in the semi-open spaces in the high-rise housing environment. Before a predictive model is developed

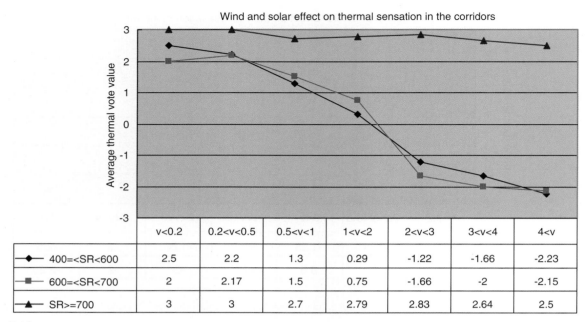

Wind and solar effect on thermal sensation in the corridors

	v<0.2	0.2<v<0.5	0.5<v<1	1<v<2	2<v<3	3<v<4	4<v
◆ 400=<SR<600	2.5	2.2	1.3	0.29	-1.22	-1.66	-2.23
■ 600=<SR<700	2	2.17	1.5	0.75	-1.66	-2	-2.15
▲ SR>=700	3	3	2.7	2.79	2.83	2.64	2.5

5.16.
Effect of wind and solar effect in the corridors, Block 1, Bedok Court. (*Source:* Q. Liang)

for semi-open spaces in the high-rise housing environment, Liang's study provides a suitable design guideline for thermal comfort for forecourts and corridors with similar climatic conditions.

5.4 PLANTS, ENVIRONMENT AND SOCIAL ACTIVITIES

Architects, urban designers, and sociologists popularly embrace the concept of gardening as a strategy towards a more humane and comfortable living environment. "Garden City", "City Beautiful Movement",[8] "Corporative Gardens",[9] "Sky-gardens" and other top down green strategies have achieved satisfactory fruits to some extent. However, some the potential owner initiated and maintained green spaces, facilitated by semi-open spaces such as the forecourt, have not received enough attention in contemporary design.

Bay (2004b) survey shows that gardening itself is one of the most popular activities in the forecourt of Bedok Court. Kong (2005) investigated the interrelationships between gardening in semi-open spaces, people and climatic performances with the case of the forecourts of Bedok Court and the corridor spaces fronting apartments of the Jurong West public housing block 510, through surveys and

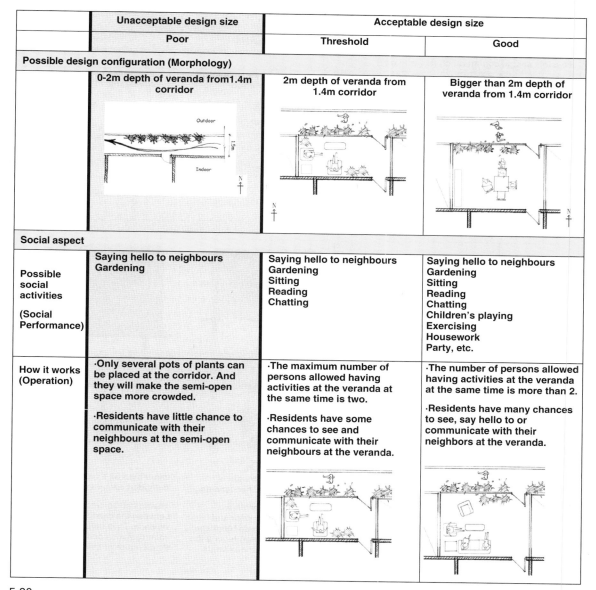

	Unaccepatable design size	Acceptable design size	
	Poor	Threshold	Good
Possible design configuration (Morphology)			
	0-2m depth of veranda from 1.4m corridor	2m depth of veranda from 1.4m corridor	Bigger than 2m depth of veranda from 1.4m corridor
Social aspect			
Possible social activities **(Social Performance)**	Saying hello to neighbours Gardening	Saying hello to neighbours Gardening Sitting Reading Chatting	Saying hello to neighbours Gardening Sitting Reading Chatting Children's playing Exercising Housework Party, etc.
How it works (Operation)	·Only several pots of plants can be placed at the corridor. And they will make the semi-open space more crowded. ·Residents have little chance to communicate with their neighbours at the semi-open space.	·The maximum number of persons allowed having activities at the veranda at the same time is two. ·Residents have some chances to see and communicate with their neighbours at the veranda.	·The number of persons allowed having activities at the veranda at the same time is more than 2. ·Residents have many chances to see, say hello to or communicate with their neighbors at the veranda.

5.20.
Summary of the guideline, morphology corresponding to social benefits and how each works. (*Source:* N. Wang & J.H. Bay)

critical, as it alerts the designer to how things work, and it is not just a matching of sizes of veranda and performances.

This set of design guidelines combines the social and environmental dynamics and thus provide an easily accessible set of "pre-parametric" knowledge for quick design decisions with regard to the design of tropical forecourts in contexts similar to those of Bedok Court.

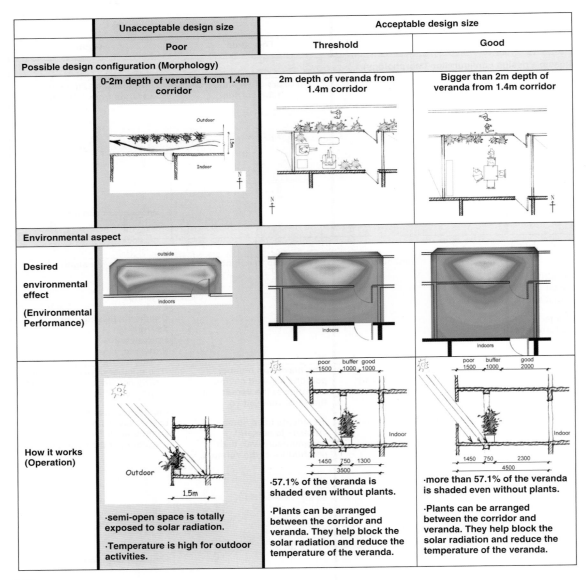

	Unacceptable design size	Acceptable design size	
	Poor	**Threshold**	**Good**
Possible design configuration (Morphology)			
	0-2m depth of veranda from 1.4m corridor	2m depth of veranda from 1.4m corridor	Bigger than 2m depth of veranda from 1.4m corridor
Environmental aspect			
Desired environmental effect (Environmental Performance)			
How it works (Operation)	·semi-open space is totally exposed to solar radiation. ·Temperature is high for outdoor activities.	·57.1% of the veranda is shaded even without plants. ·Plants can be arranged between the corridor and veranda. They help block the solar radiation and reduce the temperature of the veranda.	·more than 57.1% of the veranda is shaded even without plants. ·Plants can be arranged between the corridor and veranda. They help block the solar radiation and reduce the temperature of the veranda.

5.21.
Summary of the guideline, morphology corresponding to environmental benefits and how each works. (*Source: N. Wang & J.H. Bay*)

5.6 DISCUSSION AND CONCLUSIONS

In Singapore, it is timely that the Duxton Plain Housing International Competition organized by the Urban Redevelopment Authority in 2001–2002 encouraged a new debate on the need for community cohesion and environmental sustainability in high-rise dwelling developments in Singapore[11]. There were proposals of large sky voids, bridges and roofs, facilitating much planting as one of the strategy for sustainability, and serving as public community spaces. In all the schemes from across the world, there were no proposals of schemes similar to those of the Bedok Court forecourts that are almost more personalized and intimate socio-climatic spaces.

It is difficult but one could opt to go back to traditional lifestyle as a way to sustainability, or continue with modern lifestyle and cope with the sustainable issues with more efficient technology.Perhaps, with the forecourts, one could have the modern lifestyle, in rein-vented traditional spaces, in the high-rise apartment context and enjoy both sustainable social and environment dimensions. One can view this as a critical sustainable architecture in a similar sense of the 'critical' in *tropical critical regionalism* (Tzonis, 2001).

Bedok Court was designed to work without air-conditioning. Even though residents have fitted their own air-conditioners, they reported that they hardly turn them on, except for a few days in the hot seasons. A separate study can be made to ascertain the extent of cooling afforded by various sizes of verandas to the main building interiors.

The spatial design of the entrance forecourt can also facili-tate a synergy of socio-climatic qualities, beyond the bio-climatic. It also gives the Bedok Court a tropical and modern local character of streets in the sky. Perhaps there are other tropical architec-tural spaces, for example traditional and modern streets, which can be studied as socio-climatic phenomena to understand and hopefully achieve more critical sustainable architecture in the age of globalization.

NOTES

1 For example, Powell (1998), Lim and Tan (1998), and majority of articles on residential projects in the *Singapore Architects,* a journal by the Singapore Institute of Architects, showcase pre-dominantly low-rise living as examples of tropical urban living in Singapore. Powell, R. (1998) *The Urban Asian House: Living in Tropical Cities.* Select Books, Singapore; Lim, William S.W. and Tan, H.B. (1998) *Contemporary Vernacular: Evoking Traditions in Asian Architecture,* Select Books, Singapore.

2 The author of this paper has lived for 19 years in a simi-lar kampong from 1959 to 1978, and empathises with the experience of the architect.

3 Survey by Bay, 2000. "Design for high-rise high-density living: The tropical streets in the sky" in 21st Century QOL, Proceed-ings of the 2nd International Conference on Quality of Life in

Cities, (March 2000). School of Building & Real Estate, National University of Singapore, Singapore. And surveys in a research project by Bay in collaboration with K.P. Lam, from 2003 to 2004, entitled "Towards more robust and holistic precedent knowledge for tropical design: Semi-open spaces in high-rise residential development", National University of Singapore (Bay, 2004b). Please read this report and Bay (2005) for more details on the methods of study and results.

4 It should be noted, however, that Bedok Court is a gated condominium with controlled access, so that the sky-streets are in a sense only semi-public spaces.

5 Chua (1995), a sociologist, argues that community is formed through the increased familiarity as a result of seeing and meeting each other in everyday routine of movements and activities in and around the apartment buildings.

6 In 1986 the planning control method changed from persons per hectare to total gross area per piece of land, thus limiting Bedok Court type of developments. The author's paper (Bay, 2000) and possible incentives for more semi-open spaces were discussed at the 2nd International Conference on Quality of Life in Cities with planners from the Urban Redevelopment Authority, who were present. From then the planning regulations have been changed to encourage some semi-open spaces. Please refer to Bay (2005) for more on the restrictions and possible improvements, and differential incentives for the forecourt versus other semi-open spaces.

7 It is hardly surprising that Singapore boasts one of the highest per-capita energy consumption of any country in the world, or that one-third of all generated electricity is used for air-conditioning.

8 It was initiated since early twentieth century in U.S.A. led by Daniel Burnham. The movement inherited the idiom of European Beaux-Arts and sought to sweep away social ills through beautification. The first official expression of it was 1901 Plan for Washington, D.C.

9 Corporative gardens programs expressed differently in different states in U.S.A. since the middle of last century, such as neighbourhood corporative garden contest held in New York (1962), and Chicago (1974), and "Greening of Boston" (1987), etc. They achieved great success in revitalizing and beautifying neighbourhood.

10 Leaf area index is taken here as the single-side leaf area per unit ground area as the approximation of relative "denseness" of leaf surfaces available.

11 For example, URA, Jury Comments, News Release, 30 April 2002, report on a public debate in The Straits Times.

Sunday, May 11, 2002, and "Walking the fine line: A review of Singapore's Duxton Plain Housing Competition", by G.D. Robson and J.H. Bay (2002).

REFERENCES

Bay, J.H. (2001) Three Tropical Design Paradigms, in Tzonis, A., Lefaivre, L. and Stagno, B. (eds), *Tropical Architecture: Critical Regionalism in the Age of Globalisation*, Prince Claus Fund for Culture and Development, The Netherlands; London: Wiley-Academy, pp. 229–265.

Bay, J.H. (2004a) Socio-climatic Design in the Sky, in *Proceedings of PLEA 2004, The 21st Conference on Passive and Low Energy Architecture*, September 2004, Eindhoven University, The Netherlands.

Bay, J.H. (2004b) Research Report: Towards More Robust and Holistic Precedent Knowledge for Tropical Design: Semi-open Spaces in High-rise Residential Development. Research in collaboration with Khee-Poh Lam, Reference Number: R 295-000-034-112, Department of Architecture, National University of Singapore, 2003–2004.

Bay, J.H. (2005) Sustainable Community and Environment in Tropical Singapore High-rise Housing: The Case of Bedok Court condominium. *Architectural Research Quarterly*, arq, September. London: Cambridge.

Chen, V.F. (ed) (1998) *Architecture, The Encyclopedia of Malaysia*, Vol. 5, Editions Didier Mollet, Archipelago Press, Singapore.

Chua, B.H. (1995) A Practicable Concept of Community in High-rise, High Density Housing Environment, *Singapore Architect*, 189, 95, Singapore Institute of Architects, Singapore.

Gehl, J. (1996) *Life between Buildings: Using Public Space*, Arkitektens Forlag, Copenhagen.

Givoni, B. (1998) *Climate Considerations in Buildings and Urban Design*. New York and Singapore: Van Nostrand Reinhold.

Hagan, S. (2001) *Taking Shape: A New Contract Between Architecture and Nature*. Oxford: Architectural Press.

Hawkes, D. (1996) *The Environmental Tradition: Studies in the Architecture of Environment*. London: E & FN Spon.

HDB (2004) *HDB Annual Report 2003/04*. Housing and Development Board, Singapore.

Hyde, R. (2000) *Climate Responsive Design: A Study of Buildings in Moderate and Hot Humid Climates*, London and New York: E & FN Spon.

Jacobs, J. (1962) *The Death and Life of Great American Cities*. New York: Random House.

Kong, P. (2005) Gardening in Semi-open Spaces in Tropical High-Rise Housing: Environmental and Social Benefits, Masters of Arts (Architecture) thesis, supervised by J.H. Bay, National University of Singapore.

Liang, Q. (2005) Tropical Semi-open Entrance Space: Solar and Wind Effects on Thermal Comfort, Masters of Arts (Architecture) thesis, supervised by J.H. Bay, National University of Singapore.

Lim, J.Y. (1981) *A Comparison of the Traditional Malay House and the Modern Housing-estate House*, Unpublished dissertation, United Nations University, Tokyo.

Olgyay, V. and Olgyay, A. (1963) *Design with Climate: Bioclimatic Approach to Architectural Regionalism*. Princeton: Princeton University Press.

Ong, B.L. (2002) Green Plot Ratio: An Ecological Measure for Architecture and Urban Planning, *Landscape and Urban Planning*, 63 (2003), Elsevier Science B. V., pp. 197–211.

Robson, G.D. and Bay, J.H. (2002) Walking the Fine Line: A Review of Singapore's Duxton Plain Housing Competition, *Singapore Architects*, Singapore: Singapore Institute of Architects.

Tzonis, A. and Lefaivre, L. (2001) Tropical Critical Regionalism in Tzonis, A., Lefaivre, L. and Stagno, B. (eds), *Tropical Architecture: Critical Regionalism in the Age of Globalisation*, Prince Claus Fund for Culture and Development, The Netherlands. London: Wiley-Academy, pp. 1–13.

Tzonis, A. (1992) Huts, Ships and Bottles: Design by Analogy for Architects and/or Machines, in Cross, N., Dors, K. and Roozenburg, N. (eds), *Research in Design Thinking*, 1992, Delft: TUDelft, pp. 130–165.

URA (2004) *Residential, 4th Quarter 2004, Property Market Information*, Urban Renewal Authority of Singapore.

Wang, N. (2005) Guidelines for Socio-Climatic Design of Semi-Open Entrance Spaces of Tropical High-Rise Apartments, Masters of Arts (Architecture) thesis, supervised by J.H. Bay, National University of Singapore.

Wong, N.H., et al. (2002) Thermal Comfort Evaluation of Naturally Ventilated Public Housing in Singapore, *Building and Environment*, 37, Pergamon, Oxford, pp. 1267–1277.

Yeang, K. (1986) *Tropical Veranda City: Some Urban Design Ideas for Kuala Lumpur*. Kuala Lumpur: Asia Publications.

Yeang, K. (1996) *The Skyscraper, Bio-climatically Considered; a Design Primer*, London: Academy Editions.

6 BEAMs AND ARCHITECTURAL DESIGN IN SINGAPORE PUBLIC HOUSING

Boon-Lay Ong[†] and Chi-Nguyen Cam[‡]

[†] Department of Architecture, The National University of Singapore
[‡] RSP Architects Planners and Engineers, Singapore

Abstract

In the endeavour to achieve sustainable architecture, environmental performance – often found in Building Environmental Assessment Methods (BEAMs),
among other approaches and strategies, is increasingly and widely accepted
in contemporary practice. However, with their inherited quantitative, material-
ist and structuralist approaches, building environmental assessment methods
can conflict with other aspects of sustainable architecture – e.g. social and
economic aspects. The differences between environmental performance and
architectural approach are known among design professionals, but viewpoints
on environmental performance are diverse.

This paper aims to review critically building environmental assessment methods
from a sustainable architecture perspective, to highlight the differences between
buildings as defined by environmental performances, and by architecture as
living space. Environmental performance is then questioned as a comprehensive indicator and solution to environmental issues, through an empirical
review of the significance of socio-economic influences to actual environmental performance in the context of public housing in Singapore. The potential of
architectural approaches – the attributes that environmental performances do not
inherit – will be explored; and how building environmental assessment methods
can contribute to these potential architectural approaches will be discussed.

Keywords

Building Environmental Assessment Methods, sustainable development, housing design, socio-economic trends, Singapore public housing.

6.1   ASSESSING SUSTAINABILITY

As the need to be sustainable becomes more apparent, and more buildings lay claim to sustainability, the need to evaluate or score buildings in a numerical and objective way becomes ever more urgent. Several Buildings Environmental Assessment Methods or BEAMs are currently in use. Despite their current adoption, there are admitted problems with them. As architects, we have always argued for a broader understanding of environmental metrics (Ong and Hawkes, 1997; Ong, 1997; Cam, 2003). This broader understanding is necessary both to bring into play aspects that environmental metrics cannot take into account, and to reflect and help account for the way architects approach design. This chapter extends our argument.

To understand the underlying problems, it is useful to reflect upon the ideology behind current efforts in sustainable architecture. Hagan (2001) suggests that there are two main schools of thought: the rational and the arcadian. The arcadian approach calls for a return to nature or to a lifestyle that is in harmony with nature. The thinking is holistic. The objective is to attain a certain lifestyle that is deemed sustainable. While technical and technological solutions, like the use of recycled materials, low-energy construction and alternative energy sources are implemented, the emphasis is on choosing an alternative lifestyle that eschews many modern luxuries. Activities like growing your own vegetables and fruits, using bicycles and walking instead of driving, reducing cooling and heating bills by wearing appropriate clothing, and supporting organic and natural products are highly encouraged. If possible, living in a small town or in the country is preferred over living in the city. In this approach, the benefits gained are primarily a result of the lifestyle, and buildings designed within this approach tend towards vernacular or traditional construction and materials. They are often designed to be naturally ventilated.

The rationalist approach assumes that the modern consumer is unlikely or unwilling to change their current behaviour and consumption habits. The solution is therefore to focus on better technology and improved design. The aim is to reduce, reuse and recycle while maintaining, if not improving, current standards of living. Any benefits arising from a change in lifestyle are discounted from the rationalist's considerations. It must be said that the rationalist does not entirely dismiss lifestyle as a factor. The view, however, is that lifestyle decisions are individual choices and beyond the scope of the rationalist's influence. It is in the rationalist's approach that quantitative measures are most effective.

In practice, of course, the two positions are extremes and it is understood that both lifestyle and technology contribute to sustainability (Cam, 2004). Rudlin and Falk (1999), for example, state that "sustainable housing is defined as much by the economic forces, the social trends, and policy as by physical design". Our research presented here supports the need to look beyond the building itself for an understanding and solution to sustainability. The problem that then presents itself is how to reflect lifestyle factors in BEAMs.

6.2   A PREVIEW OF BEAMS

Over the years, a number of BEAMs have been developed. Despite their acknowledged weaknesses, these methods are helpful in many ways. They can provide an economic incentive (Prior et al., 1991); raise public awareness (Cole, 1998); act as a common means of communication in designing, constructing and sale of buildings (Cole, 1998); and identify areas of current weaknesses in design (Cam, 2003).

By definition, BEAMs are "techniques developed to specifically evaluate the performance of a building design or completed building across a broad range of environmental considerations" (Cole, 1998). Since the success of BREEAM (Building Research Establishment Environmental Assessment Method) in 1991 in the UK, BEAMs have gained momentum with the adoption of a modified BREEAM in several countries – e.g. HK-BEAM in Hong Kong and BEAM in Singapore – and the development of other methods like LEED (Leadership in Energy and Environmental Design) in the USA. These first-generation assessment methods have as their main objective the provision of an economic incentive tool to encourage developers and architects to deliver environmentally friendly buildings. The economic gains derived from BEAMs lie in the predicted energy savings, as well as in taking advantage of various tax incentives in the host countries. There is also the possibility of commanding higher sale prices as a result of the perceived savings in the long term. Being numerical, BEAMs allow developers and architects to quantify predicted savings and use these predictions as selling points. It is not yet established, however, if buildings built to BEAM standards actually result in substantial savings to the owner, and if so, how much the savings are. Our research suggests that user behaviour is a much more potent factor.

The Green Building Challenge (GBC) – a method established by an international joint effort – is a second-generation building environmental assessment method. Since it does not aim to be a commercial tool, GBC can afford to be more complex, to reflect a more comprehensive approach to building environmental assessment. As expressed by Raymond Cole (2001):

> The Green Building Challenge (GBC) process is a unique international collaborative effort that draws on the individual and collective experience of the participating countries. The process consists of the definition, structuring and scoring of a range of collectively agreed performance criteria – the GBC assessment framework, the development of a software version to operationalise the framework – GBTool, its testing on case study buildings and the presentation of the results at major conferences.

6.3 A CRITIQUE FROM AN ARCHITECTURAL PERSPECTIVE

From an architectural perspective – where the "loose fit [...] between form and performance: a space in which cultural pressures can produce strange distortions" (Maxwell's foreword in Hawkes, 1996) – BEAMs do not adequately address these extraneous factors. BEAMs assume a direct correlation between environmental problems and the technical and physical aspects of the building and ignore user behaviour. Williamson et al. (2003) commented thus:

> People are strangely absent from this image. They are assumed either as keen participants whose aims are identical ... or they are "designed" out of participation because they cannot be trusted. For example, there is a strong image that the building itself the use of energy, not its occupants. [...] This contributes to the impression that issues to do with occupancy are not a major concern when considering the environmental aspects of design.

As our research shows, "issues to do with occupancy" are indeed major concerns, if not the overriding ones. That this is so may seem obvious. No amount of design can save the energy bill of an occupant who does not practice energy-saving behaviour. The single most important influence in terms of energy consumption and waste production is surely the occupant. A side-effect of ignoring occupant's behaviour or designing them "out of participation because they cannot be trusted" is that buildings designed this way tend not to allow human intervention, and often frustrate the owners from making their own contributions to energy saving and their own adjustments to the building's performance. Conversely, and somewhat ironically, owners will then revert to energy-wasteful practices in order to achieve the level of comfort they want.

Yet, this does not preclude the importance of good design nor the use of appropriate technology. Just as a "bad" occupant will result in high energy costs, so too will bad design impede the efforts of the conscientious user. The key is to design with the user in mind. This is difficult to do, since occupant behaviour is hard to predict. A wider understanding of the various factors that influence the occupant will be useful in helping the architect identify key issues through which significant improvements in energy saving can accrue. Our research supports this approach. One of the key factors arising from our research, in determining energy consumption, is resident affluence. An understanding of the affluent lifestyle and how it can be mediated through design, and perhaps with some education through the media, will be more effective than just applying energy-saving methods to building design.

For the purpose of this chapter, however, we will discuss just two socio-economic factors and then suggest how an alternative approach might yield better results.

6.4 SOCIO-ECONOMIC FACTORS IN SINGAPORE PUBLIC HOUSING

6.4.1 Resident affluence and energy consumption

Two main factors have been identified as contributing to the success of public housing in Singapore (Wong and Yeh, 1985). First, the slum living conditions and severe housing shortage in the 1960s led to the initiation of mass public housing development. Against this backdrop, public housing was perceived clearly by its intended audience as upgrading their current living conditions. In addition, the shortage of land meant that housing is generally expensive and beyond the reach of the average young Singaporean, not just the slum dwellers and squatters. Furthermore, with compulsory land purchases by the government, private housing went further and further beyond the young Singaporean's reach. Significantly, within a few years, a waiting list was developed for public housing with a waiting period counted in years. By the 1980s, public housing had become the main residential option for the majority of the population – the exceptions being the higher-paid professionals, successful businessmen, and those who already owned a family home.

Secondly, the housing design and policies of the Housing and Development Board (HDB) have continuously evolved and transformed in response to changes in socio-economic conditions and user feedback. The government quickly saw that public housing was needed not just for the lowest income strata but for most of the population, with young newlyweds as primary targets and of greatest political concern. The provision of public housing changed from very-low-income households in the 1960s and 1970s to include middle-income households by the 1980s. The welfare of young newlyweds is of great importance politically – being young, they are at the early stages of their career and cannot afford private homes. Yet, the need for a home to start a family is great. Moreover, they will form the core workforce of society over the years, and will have the main political and social influence over their lifetime. By providing for this group, the government was able to build up a core of heartland Singaporeans. This evolution has kept the public housing stock flexible and adaptable through both good and bad times, and thus has made HDB housing the main housing form in Singapore, currently housing more than 85% of the population.

Since the 1960s, there has been increasing affluence in households in the public housing sector (Figure 6.1). As society becomes more affluent, residents' aspirations change and energy consumption increases. This happens in two ways – through owning more electrical household appliances and through the increased use of these appliances. There is a strong correlation between the trend of household affluence (Figure 6.1) and the trend of energy consumption (Figure 6.2). The influence of household affluence overrides the fact that there have been, from time to time, attempts to improve housing design in terms of environmental performance – natural ventilation, daylighting, and using low-energy lighting – with the objective of reducing domestic energy consumption in public housing (HDB annual reports 1967, 1983).

Statistical data on ownership of consumer durables by public housing flat type (Figure 6.3) indicates that there are a higher percentage of households in larger flats owning air-conditioners. Households in larger flats tend to have higher incomes (Figure 6.4) and thus are better able to afford air-conditioning compared to

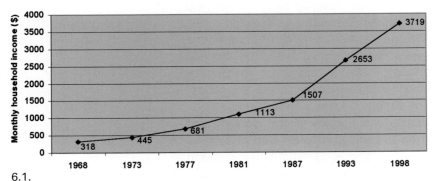

6.1.
Trend of mean monthly public housing household income. (*Source: Housing and Development Board, 2000a,b*)

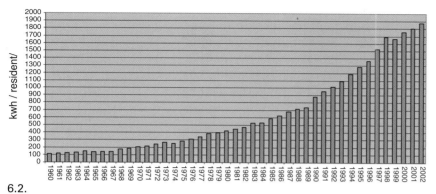

6.2.
Domestic energy consumption. (*Source: Compiled from Department of Statistics. Yearbook of Statistics, Singapore. Various issues, 1967–2002*)

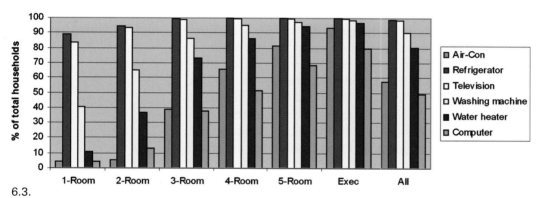

6.3.
Ownership of consumer durables by flat type. (*Source: Housing and Development Board, 2000a*)

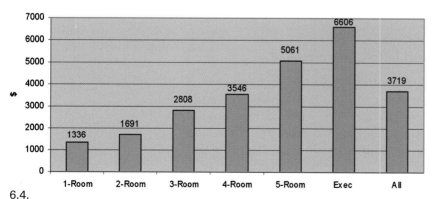

6.4.
Mean monthly household income by flat type. (*Source: Housing and Development Board, 2000a,b*)

households in smaller flats. There are also regulatory constraints by the HDB on the amount of air-conditioning allowed in flats. Figure 6.3 also shows a higher percentage of households in larger flats owning other consumer durables – washing machines, water heaters and computers. These particular appliances may be considered luxuries and lower income families will bath in cold water, not have computers at home and wash clothing by hand. The ownership of refrigerators and televisions is less pronouncedly different between the smaller households and the larger ones.

The underlying explanation for the increased ownership and use of the "luxury" items is related to the lifestyle of the families in larger flats. Note that there is a five-fold difference between the mean monthly income of one-room-flat and executive-flat owners (Figure 6.4). Both spouses in the executive-flat families are likely to be working, and so a washing machine is probably considered a necessity. Computers, as well, are probably considered necessary for the children's educational needs. Families in the smaller flats may have only one working parent, since the cost–benefit of the wife staying at home and doing the housework probably outweighs the earning capability of the wife less the cost of hiring a maid.

The figures presented here reflect the socio-economic differences between the different flat cohorts. As the size of the flat increases, the families living in them differ not just in terms of income but also in terms of family structure. Consequently, what might seem a luxury to the lower end of the households studied will be deemed a necessity to households at the upper end of the chart. To reduce the energy consumption of the families at the upper end will require attempts to reduce the energy consumption of the appliances that they use, rather than simply restricting the number of appliances they can use. An understanding of the lifestyle and socio-economic factors of the families enables the designer to identify key concerns and design specifically to address these areas.

6.4.2 Household size and land-use efficiency

The criterion regarding land resource in BEAMs suggests that a high building density is better than a low building density. The argument for this is that high-density buildings take up less ground area, reduce traveling time, and reduce the need for roads and other related amenities. High building densities imply high land-use efficiencies. However, this criterion is in conflict with other environmental criteria, e.g. daylighting, natural ventilation and privacy. A possible solution to these conflicts is to introduce a weighting system to the assessment. However, it is difficult to determine exactly how much the weighting should be introduced, and which factors are linked and in what ways to the weighted factor.

Due to the island's small size, population growth and land-use have always been major issues in Singapore. Over the years, the population has continuously increased – from about 1.6 million in 1960 to nearly 3.4 million people in 2002. Building density, as a result, has increased more than two-fold from about 2810 persons/km^2 in 1960 to 6075 persons/km^2 in 2002. Strategies to tackle the problem of land-use include reclaiming land and increasing building plot ratio.

Over the years, household sizes show a trend downwards (Figure 6.5). This further complicates the issue, as smaller households tend to consume more energy per person and take up more space per person as well. Even though there is a corresponding decrease in apartment sizes over the years, the decrease in apartment size does not compensate for the larger decrease in household size. Although not scientifically established, there is a general trend towards smaller families in cities where building densities are higher. Probably, the close proximity of others and the faster lifestyle associated with the city reduces our desire for families and greater responsibilities.

High-rise housing brings with it other problems as well. It is generally more expensive to build, requires more energy and material resources to construct, and consumes more energy per unit area during habitation. Increasing building density resolves some problems but raises new ones.

The identification of land-use efficiency as a sustainable criterion highlights the problem of separating the various factors that contribute to sustainability. Taken in itself, the argument that higher

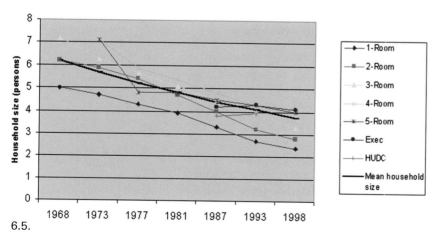

6.5.
Trend of mean household size by flat type. (*Source: Housing and Development Board, 2000a,b*)

land-use implies savings in terms of shorter travelling time, more efficient use of amenities and freeing up more land seems plausible. But increasing building density also creates problems that low land-use densities do not face. Again, a more holistic approach is needed. The advantages gained on the one hand have to be weighed against the problems raised on the other. Often, the deciding factor is not a question of sustainability but one more closely related to socio-economic considerations. The decision to go high-rise in Singapore is not based on sustainability, nor necessity, nor even economics. It has most to do with Singapore's vision of itself as a developing modern city. Once the vision takes root, the course of development will follow. It will therefore be better to develop a different, more sustainable vision than to just address sustainability issues individually as laid out in BEAMs. Architecture embodies this vision of ourselves. Our wealth is often measured by the homes we live in, the economic health of the city by state of the buildings in it, and the material development of a country by the cities and towns it has.

6.5   THE ARCHITECTURAL APPROACH

Architects practice a different approach towards sustainable development that BEAMs cannot embody. How architectural approaches can respond to socio-economic trends, particularly the two phenomena described in Section 6.4, are examined to illustrate a different sustainable paradigm other than those embedded in building environmental assessment methods. The following analysis will take examples of design efforts made in the evolution of public housing that are not recognised by the building environmental assessment method.

6.5.1   An alternative approach to energy efficiency

As shown above, residential energy consumption and social affluence are correlated. The approach of applying energy-efficiency design strategies – e.g. enhancing natural ventilation, daylighting, etc. – as found in various BEAMs is useful to some extent but is not enough to rectify the issue. The architectural approach is to identify a key issue and, in designing for that, resolve related issues as well. For example, in older public housing estates built in the 1970s, a common typology is single-loaded slab blocks arranged parallel to each other facing north–south. The design is ideal for thermal performance and natural ventilation in an equatorial location like Singapore. North- and south-facing elevations will receive the least sun. However, the potential of this orientation is not fully exploited, simply because residents often keep their windows closed to maintain privacy (Figure 6.6).

In contrast, later public housing estates are designed in a more holistic manner. The elevations are arranged to avoid windows looking into one another and to allow a pleasant view of gardens, greenery, or perhaps a distant vista (Figure 6.7). This design approach identifies view as an issue, and this is managed in terms of human satisfaction rather than sustainability or climatic considerations. Sometimes, the view demands an orientation that is not ideal climatically, and other solutions must be introduced to reduce sun insolation, for example. However, the provision of a good view encourages residents to open the windows to enjoy not only the green settings but also natural air and light, which subsequently lead to a reduction in energy consumption for lighting and air-conditioning. Coincidentally, the staggering of apartment blocks results in better ventilation as the apartments no longer block each other.

In this way, architectural design can contribute to a reduction in energy consumption without sacrificing the residents' needs and preferences. As another example, landscape is often evaluated by BEAMs as a percentage of site area, with the objective of reducing surface water run-off and urban heat island effect. In these discussions, the qualitative aspect of landscaping is absent.

6.6.
Closeness between parallel housing blocks, common in the 1970s, discourages the opening of windows. (*Source: Cam*)

6.7.
More recent housing blocks are designed with integrated greenery and with windows facing away from one another. (*Source: Cam*)

The architectural debate on landscaping includes deciding on the type of greenery to plant – e.g. more trees than shrubs, to cluster the trees together to form a naturalistic grove, to provide a lawn or to align trees of a certain species in an aesthetic manner. The different ways in which the landscape can be planted allows for or prevents certain activities. Pleasant and well-shaded landscaped public spaces provide not only environmental benefits but also an opportunity for indoor activities to flow outdoor. Indeed, the outdoor activity may not even be possible without an indoor facility to complement it. Also, a pleasant outdoor environment and atmosphere can be enjoyed indoors through open windows. With greenery outside, the window may be kept open to bring in light and cool air. With hard tarmac outside, the window is likely to be closed to reduce glare and prevent the influx of warm air and reradiated heat. The value of nature is made tangible to the residents and, at the same time, a better sustainability rating is achieved.

6.5.2 An alternative approach to land-use efficiency

The increasing pressure on scarce land resources in Singapore derives from increases in population through internal population growth, as well as increases in the number of foreign workers, increasing affluence and the trend towards smaller household sizes.

The solution to this pressure is not to build more and taller high-rise apartments but to understand the social and cultural milieu in public housing. A long-term objective of public housing is to respond to, and to promote, the traditional value of the extended family. Extended families help relieve both housing and economic pressures on the government. The advantages are numerous: from reducing the need for housing stock, to allowing both parents to work while the grandparents look after the home and children, to reducing the need for government-aided welfare for the aged. In line with this objective, the HDB began to build more multi-bedroom flats than one- and two-room flats after the 1960s. However, as society became more affluent, residents had higher demands of their living environment. New nuclear families started to form, and younger couples began to move into flats of their own (Housing and Development Board Annual Report 1970). The newer generation tends to prefer smaller families for various economic and social reasons. This has contributed to a trend toward smaller household sizes. An attempt was made to introduce a multi-generation flat type, "which allow families with grandparents or in-laws the benefits of living together under one roof whilst at the same time retaining privacy for both the younger and older generations" (Housing and Development Board Annual Report 1987/1988). However, the flats were not well received and their construction was discontinued.

Again, the underlying social considerations outweigh political and economic drivers towards a certain design direction. What the authorities did not understand was that distances in Singapore are generally small and access to family members is not a priority that requires proximity in living. Young families preferred to live a little further away, because it was not a problem to visit even on a daily basis. On the other hand, a certain distance and hence autonomy could be maintained by living further apart. Also, most families in Singapore at that time would have more than one child. This means that even if the parents were to stay with one of their children, other newlyweds would need a nuclear family home. Thus, the need for multi-bedroom flats was not sufficient to justify it as a form of public housing. A more viable, and currently practised option, is to allow extended families to buy adjoining flats and renovate accordingly.

Public housing also contributes to land-use efficiency through shared public facilities – e.g. neighbourhood parks, precinct gardens, playgrounds and markets. In older estates, these facilities are located outside the estates themselves, in a separate communal area. These facilities are integrated into the housing estates in newer housing developments – e.g. Punggol and SengKang. Housing is grouped into precincts, with the apartment blocks arranged on the periphery and a common green area or a multi-storey carpark with rooftop garden in the middle. Housing blocks can be spaced further apart, thus achieving land-use efficiency without (or with minimum) negative impacts on other environmental requirements, e.g. natural ventilation, daylighting and privacy.

Again, successfully implementing a sustainable measure depends on a deeper understanding of the built environment and how it is used.

6.6 BEAMS AND THE ARCHITECTURAL APPROACH

We have presented here the significant difference an architectural approach to design can make to attaining energy savings. Superficially, the architectural issue may not even seem to be of a sustainable nature. For example, the reason for orientation of windows is primarily to provide a nice view. However, understanding the significance and role of a good view is critical in ensuring that the windows are actually used as desired. Other aspects of design – the provision of greenery, staggering of apartment blocks – must follow from this consideration of views. If properly considered, the architectural approach leads to actual savings, because the user is more likely to act in a way that will save energy. This, however, does not undermine nor discredit the practice of BEAMs, which does at least contribute towards an understanding of quantifiable factors that contribute to sustainable architecture. First, as acknowledged by Hagan (2003), buildings with environmentally conscious design are more exposed to natural factors. These factors, with their variability and fluctuation – e.g. solar intensity, wind pattern, humidity, daylighting level, etc. – result in greater unpredictability in building performance compared to sealed buildings. BEAMs are helpful in providing a quantitative measure of building performance despite this unpredictability.

Second, with increased globalisation and mobility, user perceptions and expectations of the built environment, especially in terms of comfort level, are increasingly demanding and well-informed. The modern home-owner experiences and can compare different environmental situations both at home and abroad, leading to a demand for a higher standard of living conditions. BEAMs, as a commonly used metric, can contribute to the architect's discussion with the modern informed user.

Third, the sustainable paradigm from an architectural standpoint, using preferred qualitative measures and innovative approaches, can be enhanced and made more persuasive if there is a means to quantify and to benchmark the proposals made. It is essential in sustainable practice to have a methodology to indicate whether what we have been doing is comprehensive; and whether we are, at least potentially, on the right track (Williamson et al., 2003). BEAMs, once their role is appropriately clarified, can facilitate addressing the above concerns.

At this point, the question arises as to what is the role and appropriate application of BEAMs in delivering sustainable architecture. From a review of current literature, BEAMs are assigned many different roles, which can be extended to act as objectives or goals (Crawley and Aho, 1999) as well as guidelines (Cole, 1999) for architectural design. While we have argued that it is important to take a broader perspective in order to understand the mechanics of sustainable design, it is not possible to introduce the architectural approach as part of BEAMs, as they conflict conceptually. Architects need to study and understand BEAMs because BEAMs provide a summary of all the issues that pertain to sustainable design. However, architects need to understand their own role as being broader than the direct application of BEAMs. BEAMs serve as a backdrop as well as a final arbiter of the architect's design ideas.

As in our previous papers (Ong, 1997; Ong and Hawkes, 1997), our contention is not so much to displace environmental metrics but more to enhance the understanding of how these metrics can be applied. Properly understood, BEAMs can provide a foundation for innovative design and enhance the quality of sustainable architecture. The use of environmental metrics, often considered as hindrance to architectural design, can be instead a stimulus for innovation (Hagan, 2003). Quantitative environmental measures can provide a foundation for understanding and enriching qualitative analysis or design in architecture. Furthermore, architectural design need not engage with environmental issues per se. The consideration of other issues may not always coincide with expectations of BEAMs, but can result in measurable benefits using BEAMs. Moreover, BEAMs provide a framework for environmentally benign practice. This framework is useful as a backdrop for the designer in addressing his design problem from a fresh viewpoint. The solving of socio-economic issues can lead to good environmental practice. In this way, housing development that takes a holistic approach with innovative design is more sustainable.

REFERENCES

Cam, C.N. (2003) On Environmental Sustainability of Singaporean Public Housing Development: Reviewing from Building Environmental Assessment Methods' Perspective in *Proceedings of the 9th International Sustainable Development Research Conference*, University of Nottingham, UK, April 2003.

Cam, C.N. (2004) From Everyday Activities: A Conceptual Framework for Socio-Techno-Centric Approach to Sustainable Development. *International Journal of Technology Management and Sustainable Development* 3(1), 59–66.

Cole, J.R. (1998) Emerging Trends in Building Environmental Assessment Methods. *Building Research and Information* 26(1), 3–16.

Cole, J.R. (1999) Building Environmental Assessment Methods: Clarifying Intentions. *Building Research and Information* 27, no 4/5, pp. 230–246.

Cole, J.R. (2001) Lessons Learned, Future Directions and Issues for GBC. *Building Research and Information* 29(5), 355–373.

Crawley, D. and Aho, I. (1999) Building Environmental Assessment Methods: Applications and Development Trends. *Building Research and Information* 27, no 4/5, pp. 300–308.

Department of Statistics (annually from 1967 to 2002). *Yearbook of Statistics, Singapore*. Singapore: Department of Statistics.

Hagan, S. (2001) *Taking Shape: a New Contract between Architecture and Nature*. Oxford, Boston: Architectural Press.

Hagan, S. (2003) Five Reasons to Adopt Environmental Design. *Harvard Design Magazine*, Spring/Summer Vol. 18.

Hawkes, D. (1996) *The Environmental Tradition: Studies in the Architecture of Environment*. London: E & FN Spon.

Housing and Development Board (1967–2000) *HDB Annual Reports*. Singapore: Housing and Development Board.

Housing and Development Board (2000a) *Profile of Residents Living in HDB Flats*. Singapore: Housing and Development Board.

Housing and Development Board (2000b) *Residential Mobility and Housing Aspirations*. Singapore: Housing and Development Board.

Ong, B.L. (1997) From Homogeneity to Heterogeneity, in Clements-Croome, D. (ed), *Naturally Ventilated Buildings: Building for the Senses, the Economy and Society*, London: E & F N Spon, Chapman & Hall, pp.17–34.

Ong, B.L. and Hawkes, D.U. (1997) The Sense of Beauty: the Role of Aesthetics in Environmental Control, in Clements-Croome, D. (ed), *Naturally Ventilated Buildings: Building for the Senses, the Economy and Society*, London: E & FN Spon, Chapman & Hall, pp. 1–16.

Prior, J.J., Raw, G.J. and Charlesworth, J.L. (1991) *BREEAM/New Homes Version 3/91 – An Environmental Assessment for New Homes*. Watford: Building Research Establishment.

Rudlin, D. and Falk, N. (1999) *Building the 21st Century Home – the Sustainable Urban Neighbourhood*. Oxford: Architectural Press.

Williamson, T., Radford, A. and Bennetts, H. (2003) *Understanding Sustainable Architecture*. London and New York: Spon Press.

Wong, K.A. and Yeh, H.K.S. (1985) *Housing a Nation: 25 Years of Public Housing in Singapore*. Singapore: Maruzen Asia.

Part III

BUILDING AND PLANNING REQUIREMENTS

7 POLICY AND EVALUATION SYSTEM FOR GREEN BUILDING IN SUBTROPICAL TAIWAN

Hsien-Te Lin

Department of Architecture, National Cheng Kung University, Taiwan

Abstract

Sustainable development has been a worldwide concerned issue in many aspects. However, most of the existing Green Building assessment tools such as BREEAM, GBTool, CASBEE, and LEED are established for developed countries with cold climates, and many of them are difficult to be directly applied for those with tropical climates. For example, there are great differences between residential energy usages for the diverse climates, which may greatly influence the technology of building envelop design. Furthermore, the buoyancy ventilating method is applied well in south European countries, but is not suitable for tropical climates due to the warm outdoor temperature and high humidity. This essay will discuss different aspects of green building technologies for various climates by analysing the building energy simulation of 300 Asian cities by using two energy-consumption distribution maps in a whole Asian scale. The discussion is specifically focused on different green building design strategies such as shading, insulation and ventilation for the climate between tropical and subtropical zones. This essay will also introduce an unique Green Building Labelling System in Taiwan – EEWH system, which mainly concerns about four topics, Ecology, Energy conservation, Waste reduction and Health, comprising nine environmental indicators. This system has been maturely developed, simplified and adapted for the subtropical climate and acknowledged by the Ministry of the Interior of Taiwan as the standard method for Green Building evaluation since 1999. From 2005, a new rating scheme will be brought in the EEWH system. The ratings are classified into four levels ranking from diamond, gold, and silver to bronze, which respectively represent the top 5, 15, 30, and 50% of the score percentage. In the near future, the EEWH system with new rating scheme will become a more important index in a government-led program encompassing long and lasting development in the technology and promotion of green buildings. Finally, some Green Building promotion programs, such as mandatory Green Building Green policy, Green Remodelling project, Green Building Award competition, and Green Building Expo will also be introduced in this essay.

Keywords

EEWH, Green Building, rating system, subtropical climate, building energy.

7.1 LOCALIZATION OF GREEN BUILDING POLICY IN TROPICAL CLIMATES

7.1.1 Local action of Green Building assessment

Since the 1992 Rio Earth Summit, sustainable development has been an issue of worldwide concern in many aspects. In architecture field, the sustainable development movement has been gathered as a main trend and the assessment method of Green Building became a fashionable field of research throughout the world. After the first Green Building assessment tool BREEAM in the world was proposed by BRE (Building Research Establishment) in Britain 1990, more than 14 countries have established their own assessment tools such as GBTool (Canada), CASBEE (Japan), LEED (U.S.), EEWH (Taiwan), within ten years.

To catch up with the sustainable building trend, many countries tend to directly transfer or modify the existing foreign Green Building evaluation tool to their domestic application. "Think globally, Act Locally" is well represented in sustainable building field, but it is very difficult to be applicable because it requires a very long term research for localization. Such localization is not just a modification process from foreign tool but to establish the independent categories, index, standard, weighting system, and rating method according to the domestic climatic and environmental conditions. Sometimes, the transformation or modification of foreign tool without sophisticated and careful localization may create a great obstacle to the future development of Green Building policy, especially in the modernization process of developing countries. Most of the existing evaluation tools of Green Building were established at developed countries with cold climates, and many of their green technologies are difficult to be directly applied to those with tropical or subtropical climates. Furthermore, the great differences of social background such as energy structure, building industry ... may decrease the appropriateness and reliability of foreign evaluation tool and may cause an unpredictable distorted result when a mandatory policy is based on the tool.

For example, there are great differences between residential energy usages in diverse climates, such as the domestic energy usage shared 50% in middle Europe, 37% in U.S., 26% in Japan (1999), 18% in Taiwan (2000), and 20% in Singapore (2003) which may greatly influence the technology of building envelop design. Such an energy structure demonstrates that the domestic energy conservation technology should be more efficient in the cold climate than in the tropical or subtropical climate. At the same time, the clean hydropower shared 99, 30, 14, 5, and 0% in Norway, Canada,

Japan, Taiwan, and Singapore, relatively in the electricity structure of 2002. It means the deviation in the evaluation of CO_2 emission of electricity consumption of buildings may change from 5 to 100 times between different countries and it is obviously very dangerous to adopt the foreign Green Building tool without appropriate localization.

Furthermore, due to the great difference of environmental problems for different countries, the Green Building tool is supposed to change or adjust the structure of assessment indicators. For example, the immense market for reinforced concrete construction is one of the biggest environmental killers in Taiwan. Different from the great market share of steel and wooden buildings in the U.S., E.U., or Japan, more than 95% of the building market is shared by reinforced concrete in these 10 years and makes Taiwan the second largest consumer of cement in the world in 2000. This great concrete market was primarily caused by the low priced illegally excavated aggregate (sand and stone for concrete). Despite a real aggregate market of 110 million tons per year, legal excavation of aggregate can only supply 42% (46 million tons) of the actual required amount. This means 58% of the aggregate is taken from illegally excavated sources and 80% of the illegal aggregate is reported to be excavated from rivers. This great concrete market has brought about many disasters, such as flooding, landslide, and bridge collapses which occur more and more frequently (Figure 7.1). This phenomenon has forced the Green Building policy to reduce the usage of cement and encourage steel and wooden design in Taiwan.

7.1.
A dangerous bridge with footings exposed due to the illegal excavation of aggregate from rivers. (*Source:* Author)

On the other hand, brick structure building still occupy 95% building construction market for a long time in China and has become a great environmental problem in Chinese Green Building policy. According to the national construction department of China, a brick structure apartment building with 10 000 m^2 floor area should consume two million pieces of brick and destroy 0.22 ha of farm field. Until 1999, there are 120 000 brick kilns around China which produce 600 billion pieces of brick and destroy 80 000 ha of farm every year. To decrease the huge damage from brick structure building market, the Chinese government instituted a regulation to prohibit the construction of solid clay brick buildings and the reinforced encourage concrete or the hollow brick construction at 160 major cities of coastal areas from 30 June, 2003. It is very interesting to find the dilemma for adopting reinforced concrete structure in Taiwan and China and to prove that a misuse of foreign Green Building tool may bring about a wrong building policy, even unpredictable environmental disasters.

7.1.2 Efficiency of insulation and shading in Asian climate

The first important task in the localization of Green Building tool is to establish a localized evaluation system of building energy. In order to calibrate a localized energy design tool, it is necessary to grasp the relative weightings according to the climatic contexts and building energy usages, especially between the energy efficiency and technology cost. Basically, insulation and shading are two major factors in relation to the energy design of building envelope. The efficiency of insulation increases linearly according to the degree days which represents the accumulated indoor–outdoor temperature difference. On the other hand, the efficiency of shading device shows a close relationship to the solar radiation among high temperature seasons. It is very important and difficult to choose an optimal combination of insulation and shading strategies according to its climatic context.

In order to realize the different aspects of Green Building technologies for various climate, Lin has demonstrated the relative efficiency of insulation and shading by analysing the building energy simulation of 300 Asian cities using two energy consumption distribution maps for one typical residential house and one typical office building space in a whole Asian scale as shown in Figure 7.2. The isolines of annual total thermal load of building envelope on these maps are drawn by the simulated annual cooling load and annual heating load which are predicted by Lin's simplified method (Lin, 1985, 1987) based on monthly weather

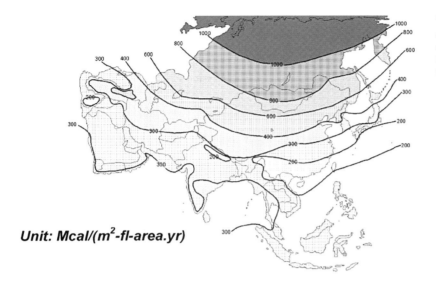

7.2.
An annual thermal load distribution map for a typical residential building in Asia. (*Source:* Lin, 1985)

Unit: Mcal/(m²-fl-area.yr)

data of maximum temperature, minimum temperature, and solar radiation in 300 weather stations across Asia. By comparing the distribution tendency of annual thermal load in such a global scale, we can see the relative efficiency of energy design technologies and find an optimal design method under different climatic context. Our comparison of these two maps led to the following investigations:

1. Annual cooling load increases gradually to the tropical climate and the annual heating load increases gradually to the polar climate, there should be an optimal region of building energy consumption which is the annual total thermal load or the sum of annual cooling load and annual heating load as shown in Figure 7.2. Colder climates in northern region of this optimal region and warmer climates in the south of this region can be called as "the insulation priority region" and "the shading priority region" relatively, with its efficiency of insulation or shading increases significantly from this region to northern and southern directions. At the same time, this optimal region can be named as "the hybrid region of insulation and shading" due to low efficiency in both insulation and shading.

2. The optimal energy consumption regions, as shown in Figure 7.2, for this simulated house with 8 mm single glass glazed area ratio 30%, no outdoor shading and poor insulation, are distributed around the subtropical belt from northern Turkey, southern Nepal, southern China, Taiwan, and southern Japan. Colder or warmer area outside this belt, the annual thermal load becomes larger due to the increase of the cooling or heating

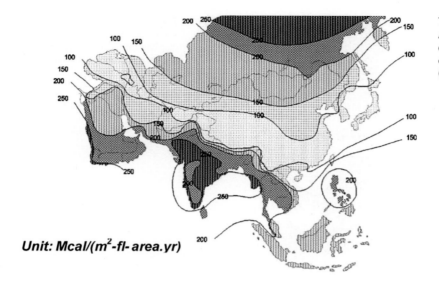

Unit: Mcal/(m²-fl-area.yr)

7.3.
An annual thermal load distribution map for a typical office building space in Asia. (*Source:* Lin, 1985)

load is larger than the decrease of the heating or cooling load. It means that the subtropical climate might have a minimum energy requirement in the world and present the typical climatic context of "hybrid region of insulation and shading" for residential building design.

3. As shown in Figure 7.3, the optimal energy consumption regions for this simulated office building space, with 15-cm reinforced concrete wall, south orientation, 10-mm single glass glazed area ratio 25%, are located in the moderate belt between latitudes 25°N and 40°N. These regions are distributed in colder climates to the north of the optimal energy consumption regions of residential building for greater cooling energy requirement in office buildings than in residential buildings. This fact tells us that energy saving efficiency of outdoor shading can be maintained from the equator to subtropical belt for residential buildings and can be even expanded to colder climates until latitudes 40°N for office buildings due to higher cooling energy consumption.

4. In tropical regions, the efficiency of insulation may be a little greater than in the optimal regions of office buildings, but the shading efficiency is absolutely greater than the insulation efficiency. It is obviously more economic to reduce the solar heat gain by using deep verandas or sun break devices than to insulate the envelope by using outer wall insulation or double skins of glass curtain wall.

The above analyses point out "insulation priority" and "shading priority" as the main climatic design strategies for colder climates and warmer climates. However, the glass box design of building skin with low shading performance, a symbol of the modern European fashion, spreads all over the tropical regions and exhausts large amount of energy with escalating speed. It is believed that the popularization of glass box buildings is the major reason of urban energy crisis in most cities of hot climates. "Glass and Steel" has been considered as the key to modern architecture in European world but it became a climatic confusion of building design in tropical countries.

Generally speaking, glass has a very low thermal conductivity, which equals to 1/600 of silver, 1/200 of aluminium, 1/50 of steel, and 1/2 of concrete, it is absolutely an excellent material for insulating the temperature difference but very inadequate for shading design due to its high solar heat gain coefficient. This characteristic makes glass very difficult to be designed as a transparent skin without huge cooling energy consumption. The localization of Green Building in tropical climates has to give up the prejudice of European aesthetics of glass box design at first and establish a sustainable identification for the tropical vocabularies of verandas, sun breaks, moderate openings, and deep shadings.

7.1.3 The climatic expression of ventilating design in tropical regions

In addition to designs of insulation and shading, the ventilating design is the third important factor for shaping a climatic form of Green Building. However, the efficiencies and technologies of ventilation design are quite different with various weather conditions and building functions, such as temperature, wind velocity, humidity and building space organization. There are two major principles in the ventilation design, cross ventilation and buoyancy ventilation, which should be kept in designer's mind in order to create a ventilation orientated building form. For cross-ventilation to utilize wind pressure, it can only be applied in the space with air current of considerable velocity. On the other hand, buoyancy ventilation is created by the indoor–outdoor air–temperature difference and can only be utilized in the climate with cooler outdoor air temperature than in the indoor comfortable zone.

As shown in Table 7.1, Lin (2004, p. 188) simulated the VUP (Ventilation Utilization Possibility), the possible ventilated time divided by the total calculated time, to investigate the potential for cross ventilation and buoyancy ventilation in various climates of the world. We find that the buoyancy ventilation design is very suitable to be applied in those temperate climates with high potentiality of

Table 7.1. Ventilation Utilization Possibility (%) in various climates

Climatic zone	City (Country)	VUP of Cross Ventilation	VUP of Buoyancy Ventilation	Total VUP
Cold or Temperate climates	Boston (U.S.)	1.86	5.21	7.07
	Portland (U.S.)	1.77	9.76	11.53
	San Francisco (U.S.)	0.79	12.27	12.93
	Sapporo (Japan)	5.53	15.26	20.79
	Los Angels (U.S.)	3.15	33.37	36.52
	San Diego (U.S.)	7.81	31.14	38.95
	London (U.K.)	2.75	28.32	31.07
	Rome (Italy)	11.79	23.23	35.02
	Beijing (China)	14.71	14.36	28.53
	Tokyo (Japan)	12.92	15.98	28.90
Subtropical Climates	Tampa (U.S.)	15.99	11.05	27.04
	Naha (Japan)	13.30	7.23	20.53
	Shanghai (China)	10.73	15.15	25.88
	Nanjing (China)	12.96	14.80	27.76
	Hongkong (China)	21.92	13.07	34.99
	Taipei (Taiwan)	25.26	15.98	41.24
Tropical Climates	Manila (Philippine)	45.08	0.63	45.71
	Singapore	26.18	0.0	26.18
	Kuala Lumpur (Malaysia)	40.11	0.0	40.11
	Jakarta (Indonesia)	34.86	0.03	34.90
Hot-dry Climates	Phoenix (U.S.)	5.39	15.46	20.84
	Salt Lake (U.S.)	1.49	6.02	7.48

Note: The data of VUP is hourly calculated from the Average Year Weather Data for the cities in Taiwan and Japan and from TMY2 for the other cities. The calculation condition are: temperature 20–30°C, wind velocity 0–3.0 m/sec, relative humidity 40–90% for cross ventilation calculation, temperature 12–20°C, wind velocity 0–3.0 m/sec, relative humidity 0–100% for buoyancy ventilation calculation.

23–34% VUP of buoyancy ventilation, such as at Los Angels (U.S.), San Diego (U.S.), London (U.K.), Rome (Italy), but is inefficient to be utilized in the tropical regions with very low VUP (near to zero). This analysis proves that the buoyancy ventilation technologies, such as the ventilating tower, roof ventilator, and high inner courtyard used in Queens Building of De Montfort University (Leichester, U.K.) in Figure 7.4, are applied very well in some temperate climates like south European countries but is absolutely not suitable for tropical climates due to the warm outdoor temperature and high humidity.

Furthermore, we can find that the areas of tropical climates, such as Singapore, Manila, Kuala Lumpur, and Jakarta, are the best regions in the world for cross ventilation utilization with 26–45% VUP, besides, hot-dry climates and cold climates are unsuitable for cross ventilation design, for it might be too dry or too cold.

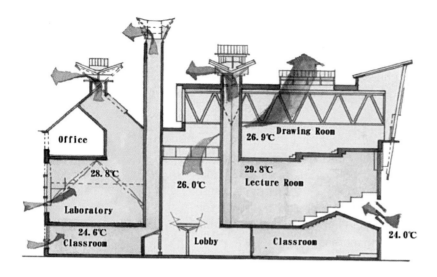

7.4.
Buoyancy ventilation concept for Queens Building. (redrawn after Steele, J., 1977, p. 65)

This fact tells us that the cross ventilation is a suitable resource for sustainable design of tropical architecture. Therefore, there still exists a very high potentiality of cross ventilation design in tropical climates especially for some intermittent air-conditioned buildings, such as residential building, sanatorium, dormitory, and school. The designer should not give up the tropical vocabularies, rectangular layout, short indoor depth, two side opening, veranda, and deep shading roof, to sculpture a sustainable tropical building form.

7.2 A GREEN BUILDING ASSESSMENT TOOL IN SUBTROPICAL TAIWAN

7.2.1 The confusion of architectural style in subtropical climate

Compared to the clear characteristic of "insulation priority" and "shading priority" in cold climates and tropical climates, the Green Building expression in the subtropical climate is more complicated and ambiguous. As mentioned in Section 7.1.2, the subtropical climate region requires minimum cooling and heating energy and presents the confusion context of "hybrid region of insulation and shading". At the same time, the subtropical climates possess medium efficiency of cross ventilation and buoyancy ventilation as shown in Table 7.1, it can also be described as "the hybrid region of cross and buoyancy ventilating technologies", for its medium potentiality of ventilating and a mixed expression of ventilating languages.

As its hybrid characteristic between hot and cold climates, cooling and heating requirements, insulation and shading efficiencies, and between ventilation and buoyancy ventilating technologies, the sustainable form of subtropical architecture might always appear to be with diversified and ambiguous climatic languages. For some enthusiastic researchers who are eager to establish a clear, solid vernacular form for subtropical architecture should notice this hybrid architectural characteristic.

To represent an adequate subtropical Green Building policy, the following is the one developed in the subtropical Taiwan. For adequate building style of Taiwan, Lin (2004, p. 223) investigated the energy performance of building envelop design by using the dynamic DOE2.0 simulation as shown in Figure 7.5. We found that energy consumption of air conditioning increases about 1.0% when the glass ratio of office building envelope increases every 1.0% with any kind of glass material in subtropical Taiwan.

This analyses proved that the glass box design of building envelop is by no means an energy killer not only in tropical region but also in subtropical climates. Due to the great solar heat gain of glass skin, the transparent dream of building skin may be suitable in cold or moderate climates for the excellent insulation performance of glass, but will not be environmental friendly in tropical/subtropical climates. Many designers are eager to find innovative glass technologies, such as reflective glass, double skin glass, and low-E glass, to cut down the cooling load of glass skin but still cannot change the basic material performance that bigger solar heat

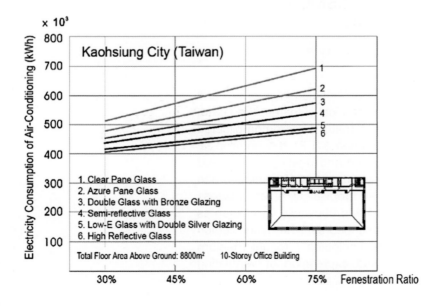

7.5.
Relation between annual air conditioning load and fenestration ratio for a 10-storey office building in a southern tropical city of Taiwan. (*Source: Lin, 2004*)

gain comes from a bigger glass opening. Shading design with veranda, sun break, and appropriate opening, is always more efficient and cheaper than any glass technology for climatic control in subtropical regions. These building languages have become the main energy design methodology of the Green Building assessment in subtropical Taiwan.

7.2.2 Background of Green Building policy in Taiwan

The building industry of Taiwan today is confronted with several crucial environmental problems and sustainability issues. First of all, recent studies show that the climates in the cities of Taiwan are getting warmer and warmer due to lack of adequate policies for urban environment and building industry in the past. Overcrowded urban population, shortage of green spaces, impermeability of the living environment, and inefficient building design for energy consumption have all caused serious climatic problems.

The higher temperature of the urban environment significantly aggravates cooling energy consumption and carbon dioxide emissions, and accelerates green house effect in the cities. The temperature differences between downtowns and the suburbs at most metropolitan areas in Taiwan are 3–4°C during peak summer months, as shown in Figure 7.6. According to Taiwan Power Company's report, air conditioning electricity consumption increases about 6% when the outdoor air temperature increases 1°C. It indicates that the energy consumption of cooling is about one-fourth higher in downtown than in the suburban areas in summer. Moreover, Taiwan is highly dependent on imported energy, with a percentage over 98% in 2003. The building industry accounts for 28.3 percent of the nation's total energy consumption (including building material production 9.77%, construction transportation 0.53%, housing energy 12%, commercial energy 6%) in Taiwan (ABRI, 2001). Therefore, the Green Building evaluation system and policy need to be incorporated with proper building energy saving techniques and regulations.

Second, Taiwan receives abundant rainfall with an average annual precipitation of more than 2 500 mm. The amount of water per capita, however, only reaches one-sixth of the world's average. Therefore, from water resource perspectives, the Green Building policy and its related strategies in Taiwan should focus on water conservation and reuse issues. Third, the immense market of reinforced concrete building, which constitutes more than 95% of the entire construction market, is one of the biggest environmental killers in Taiwan. Reinforced concrete construction is typically

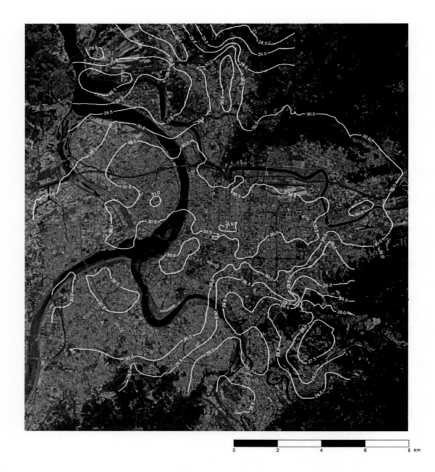

From metropolitan region's mid-night heat island temperature distribution diagram, it is clearly seen that high temperature covers the basin center of Taipei city. (Field recorded map made by Siraya Laboratory, Department of Architecture, National Cheng-Kung University at 02:00 on July 24, 1998, where the red spot is the field recording 640 points.)

considered as a high-polluted building method for its enormous energy and resource consumption. The large amount of cement used also results in the generation of 11 million tons of construction wastes per year.

Last but not least, a deteriorating environment damages people's health. Among the top ten causes of death in Taiwan, cancer and respiratory diseases share a relatively high proportion which is more than 30%. In spring 2003, the death toll due to SARS outbreak reached 73 and seriously harmed the economy of this island. A place with good indoor environmental quality has become one of the major concerns for people's daily activities.

With respect to the evaluation system for building environments, various environmental assessment methods have been proposed over the last decade. The compatibility of these systems that are developed in temperate or frigid regions with subtropical Taiwan, however, needs to be investigated. The evaluation system for Green Buildings in Taiwan should be simplified and localized, to accommodate the climatic characteristics (high humidity and high temperature), and correspond to Taiwan's local environmental issues.

7.2.3   EEWH system in subtropical Taiwan

In order to examine the environmental performance of the building, an appropriate evaluation system that is accommodating the subtropical climate in Taiwan should be developed. The ABRI (Architecture and Building Research Institute) came up with an evaluation system for Green Buildings that was first announced in 1998 and comprised of seven evaluation indicators, greenery, soil water content, energy conservation, water conservation, CO_2 emission, waste reduction, and sewer and garbage improvement. Over the past few years, the ABRI modified the evaluation system by further introducing two new indicators to the original ones, Biodiversity and Indoor Environment Quality. An evaluation system, integrated with nine categories, was thus established in 2003, as listed in Table 7.2. These indicators can be divided into four categories, ecology, energy conservation, waste reduction, and health (known as EEWH system). Quantitative indicators and the corresponding criteria for each category are described as follows.

Table 7.2. Categories, indicators and factors for Green Building evaluation in Taiwan

Categories	Indicators	Climate	Biology	Water	Soil	Energy	Materials	Evaluation factors and units
Ecology	1. Biodiversity	*	*	*	*			Biotope, green network system
	2. Greenery	*	*	*	*		*	CO_2 absorption (CO_2-kg/m^2)
	3. Soil water content	*	*	*	*			Water contentment of the site
Energy conservation	4. Energy conservation	*				*		ENVLOAD, AWSG, Req, PACS, energy saving techniques
Waste reduction	5. CO_2 emission	*				*	*	CO_2 emission of building materials (CO_2-kg/m^2)
	6. Waste reduction					*	*	Waste of building demolition (−)
Health	7. Indoor environment				*	*	*	Ventilation, daylight, noise control, eco-material
	8. Water conservation			*				Water usage(L/person), water saving hygienic instrument (−)
	9. Sewer and garbage			*			*	Sewer plumbing, sanitary condition for garbage gathering

Ecology in EEWH

Biodiversity, greenery and soil water content are three indicators for the category of ecology in EEWH. The evaluation items in biodiversity indicator include ecological network, biological habitats, plant diversity, and soil ecology. Since the purpose of this indicator is to protect biodiversity and environmental balance in a large-scale ecosystem aspect, the indicator is not applied to the site that is less than two acres. The indicator of greenery introduces the CO_2 absorption factor as the conversion unit for different types of plantings, such as trees, shrubs, climbers, etc. As a qualified green design for Green Building, the total CO_2 planting absorption should reach certain high level, with a planting rate of more than 50% of its open space and a CO_2 absorption efficiency higher than 600 kg-CO_2/(m2.40 y). The soil water content indicator is introduced to maintain the site of high permeability performance. An index for the permeable ratio of a constructed site in comparison with a bare site is adopted to evaluate the water content capacity of the site. The calculation of the permeable ratio is expected to encourage permeable pavement, ponds, permeable lowlands, and gardens on impermeable floors or roofs in the site design. A building project can be qualified as a Green Building if the permeable area is more than 80% of its open space.

Energy

The energy conservation indicator is the most sophisticated field and most localized tool in EEWH system. This indicator mainly focuses on the energy performance of the building envelope, air conditioning and lighting, which share over 80% of the total building energy consumption in Taiwan. The building envelope energy performance evaluation is quite convincing due to the building energy index, ENVLOAD (thermal load of envelope for air conditioning buildings), AWSG (average window solar heat gain for schools or huge space buildings) and Req (equivalent window ratio for residential buildings), have been involved in the building code of Taiwan since 1995. These indices especially emphasize the technologies of shading device and prohibit excessive window opening so as to achieve tropical/subtropical building expressions. The PACS (Performance of Air Conditioning System) method is also well established in the air-conditioning field in Taiwan to prevent the over design of heat source capacity, encourage the design of high efficient chillers and innovative energy saving technologies. The lighting energy can easily be evaluated based on the average efficiency index of illuminations. In the Green Building evaluation, the energy saving rates for these three indices, the building envelope energy performance, air conditioning, and lighting, should be greater than 30% of the average energy consumption.

Waste reduction in EEWH

CO_2 emissions and waste reduction are the two indicators in the category of Waste Reduction in EEWH. The CO_2 emissions indicator is an important tool for reducing pollution emissions through building materials selection and construction methods. A Green Building project should emit 18% lower than the average CO_2 emissions from typical reinforced concrete buildings through a more logical and efficient way of design in structural system and low energy materials selection. This evaluation can greatly encourage lower environmental impact structures, such as lightweight steel structured buildings, industrialized construction methods or wooden buildings. The waste reduction indicator is utilized in evaluating solid waste and particle pollution, from basement excavation, construction, to destruction in the life cycle of the building. A certified Green Building project is required to cut 10% of the soil waste, construction waste, destruction waste, and to reduce 40% of the construction particles, in comparison with the average waste emissions from reinforced concrete buildings. This evaluation can encourage more natural site design with fewer landscape alterations, less basement excavation, and low pollution construction, such as industrialized building methods and steel or wooden buildings. Recycled materials, such as recycled blocks, tiles, aggregate, are particularly encouraged in the destruction waste evaluation.

Health in EEWH

Indoor environment quality, water conservation, and sewer and garbage are the three indicators for the category of Health in EEWH. The indoor environment quality indicator focuses on the evaluation concerning building acoustic environments, lighting and ventilation environments, as well as building materials. The indicator also encourages the utilization of Green Building materials, which are natural, ecological, and recycled. The criteria are developed based on an expert system approach, and the summation of the score of each evaluation item multiplied by its corresponding weight factor can easily be calculated. The water conservation indicator is aimed at saving water resources. Many types of water saving hygienic instruments, such as water closets, bathtubs, showers, etc., are encouraged in the evaluation. Water recycling systems for wastewater or rainwater are greatly encouraged in the assessment as well. The sewer and garbage indicator does not evaluate sewer and garbage biotechnology but focuses on the landscape design and detailed improvements for sewer plumbing and garbage holding area sanitary conditions.

The above mentioned nine indicators in four categories are evaluated independently to respond to the various environmental impacts upon the earth. Each indicator has some quantitative calculation methods, equations and criteria for the evaluation process. This system has been simplified, quantified, and localized for the subtropical climate of Taiwan and is regarded as a standard evaluation method for Green Buildings by the Taiwan Government.

7.2.4　Rating system for EEWH

With this EEWH since 1999, the ABRI announced the Green Building Evaluation Handbook and Green Building Logo, which are the evaluation tool and identification marking for Green Buildings. At the same time, "Green Buildings Committee" was established to carry out the selection and certification of Green Buildings, giving great impetus to Green Building movement. Every existing building and every new building scheme before construction is encouraged to pursue the Green Building Logo shown in Figure 7.7. Based on this system, Taiwan government acted very aggressively to demand all the governmental buildings to pass the evaluation of this system since 2001. Under such encouraging circumstance, about 500

7.7.
Green Building logo in Taiwan.

綠建築標章
GREEN BUILDING

newly designed building projects had already passed the Green Building Evaluation towards the end of 2004.

However, due to the compulsive policy of Green Building Evaluation on governmental buildings, the passing standard had to be set at a relatively low level so as not to become big obstacles to the schedule keeping of public construction. The passing standard for Green Building Evaluation was required at a basic level and about 85% of the previous qualified projects have passed at relatively low scores of passing line. This kind of compulsive policy and low standard of evaluation has become an obstacle to the promotion policy of Green Building.

In order to resolve this problem, Lin (2005) has established a new rating system based on the analysis on the previous 185 qualified Green Building projects. This new rating system is keeping the same quantitative indices, criteria of evaluation and utilizes a new scoring system for nine indicators according to a normal scoring distribution of the mentioned 185 projects. The new rating system, which is reconstructed on a hypothesis of logarithm normal distribution, has created four labelling rankings which are diamond, gold, silver, and bronze as shown in Figure 7.8. Their scoring probabilities are of top 5, 15, 30, 50%, and with the scores of 53, 43, 37, 31. The new rating system is designed to avoid the low level standard in the old system and can act as an efficient assessment and promotion tool for the Green Building policy in Taiwan.

However, no matter how sophisticated the evaluation system is, it cannot cover all aspects of the green technology, so we have to be flexible for evaluating unknown innovation of Green Building. To promote the innovative design of green technologies, the new EEWH system creates an extra encourage scoring method to compensate outstanding design which cannot be evaluated by the existing indicators and categories. This method can give an extra

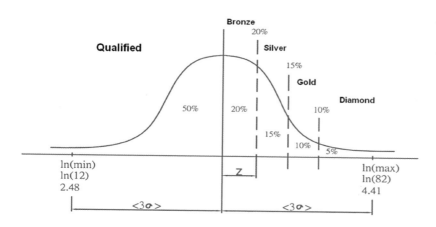

7.8.
New rating rankings of EEWH.
(*Source:* Lin, 2005, p. 135)

10–50% to each scoring of category according to the Green Building Committee based on the proposed report or explanation of the innovative design by the designer. However, this promotion mechanism can be approved only for qualified Green Building projects of EEWH and on unique ideas or technologies with close relation to the four categories of ecology, energy conservation, waste reduction, and health.

7.2.5   Green Building promotion program in Taiwan

In order to resolve the environmental problems and mitigate the impacts of urban development and building construction, the ABRI has proposed a Green Building Promotion Program. This program essentially forges a comprehensive mechanism to provide resources, research, guidance, training, and education to support the development of Green Building in Taiwan. The contents of the Green Building Promotion Program can be summarized as follows.

Mandatory Green Building design for new governmental buildings

For initiating the development of Green Building, the government of Taiwan realized that the mandatory policies can play an absolutely important role. Therefore, the Green Building Promotion Program was authorized directly by the Executive Council and the Green Building Labelling process was raised as the National identification system since the Program started. One of the major promotional strategies in this program, which also runs first in the world, is to initiate mandatory Green Building design for governmental buildings. Purpose of this strategy is to initiate Green Building design with public buildings and to encourage private sectors to adopt the Green Building concepts, so as to gradually evolve a mechanism for the entire building industry. Every governmental building project with construction cost more than 1.5 million US dollars would not be issued the building license unless it received the Green Building Certificate first. The ABRI established a Commission on Green Building that includes three committees and forty members to ensure the effective evaluation for Green Building. The commission also provides technical consultation during the design process.

Green remodelling projects

The ABRI also subsidized green remodelling projects for existing buildings. From 2002 to 2004, ninety green remodelling projects were completed for official buildings and public schools, with the total budget reaching 19.29 million US dollars. The green remodelling projects include (1) the improvement of soil water content, (2) creation of ecological biotope, (3) building envelope improvement, and (4) HVAC improvement.

In 2003, forty-eight green remodelling projects were improved for official buildings and public schools. In the improvement projects of soil water content, one water permeable grass gutter, 14 water permeable pavements, and one infiltration pond were constructed to achieve urban water cycle. Figure 7.9 is one example of permeable pavement project at Conscription Agency where the original impermeable car park was replaced with permeable cement blocks and recycled aggregate. This program achieved the biodiversity improvement by establishing one ecological pond, one ecological biotope, one ecological forest, and 13 green fields. The ecological biotope project, including the constructed wetlands combined with the function of wastewater treatment and reuse, and so forth. Figure 7.10 shows a successful remodelling project of the constructed wetland integrated with wastewater treatment.

The sun-shading devices can effectively conserve energy and exhibit the architectural style of hot and humid climate, therefore the modification of the sun-shading boards is the most important part of this program. In 2003, there are 14 cases of sun-shading device installation, and sums up to a total area of 3297 m^2 of metal shading boards and 3246 m^2 of shutter-style shading devices. We pay special attention to the harmony of the original building form and

7.9.
An example of permeable pavement improvement project in Green Remodelling Project, 2003. (*Source:* Author)

7.10.
The constructed wetland in the Architecture Department of NCKU is a veritable demonstration of ecological remodelling. (*Source:* Author)

adding the sun-shading device. Ethereal porous metal sun-shadings are adopted to enhance the building's appearance in remodelling project like Taitong High School (Figure 7.11) or Tainan Teachers' College (Figure 7.12). These shadings not only improve the quality of the indoor lighting, the building aesthetic, but also exhibit a vernacular architectural style for hot and humid climate. In the HVAC improvement projects, the ABRI selected twenty-eight buildings with air-conditioning improvement potentials in 2003. The procedure is mainly to diagnoses the HVAC performance of existing systems, and then to conduct the energy simulation of HVAC load to find the optimal operation strategy. The energy saving efficiency is required to be more than 40% in the selected projects.

Activities for Green Building education

The ABRI promotes the green building concept through a series of activities, including seminars, training courses, conferences, technical tours, Green Building Award competition, and Green Building Expo. Some technical tours and seminar courses of Green Building projects are prepared for students, teachers, architects, engineers, and governmental officers in building industry related fields. Furthermore, the Green Building Award competition was held in 2003 and 2004 in order to encourage the innovative technologies of Green Building and increase the motivation of Green Building design.

7.11.
The sub-tropical style of the modified façade; the color of the shadings is in harmony with the original color of the Taitong High School. (*Source:* Author)

7.12.
The opening of a building of Tainan Teachers College is installed with aesthetic, light-weighted, metal shadings. (*Source:* Author)

7.13.
228-Memorial hall in Chayi, Green Building Award project, 2003. (*Source:* Author)

Figures 7.13 and 7.14 show two Green Building Award winners in 2003, and all these outstanding designs are especially well coordinated with the subtropical/tropical climate or local aboriginal culture. Besides, the Green Building Expo in Taiwan was held from January to March in 2004. More than 100 000 people visited the Expo and the Green Building concept was extensively promoted to the public. All of these incentive efforts were expected to efficiently stimulate the Green Building market in Taiwan.

7.14.
An office building of Taiwan Power in Shinying, Green Building Award project, 2003. (*Source:* Author)

7.3 CONCLUSIONS

The EEWH system and the Green Building development of Taiwan described in this chapter is a unique experience for a tropical/subtropical country to catch up with the sustainable building movement in Europe. In a certain meaning, the localization of Green Building policy means a self identification process of technologies, aesthetics and architectural culture. The scientific approach with sophisticated research of climatic context and environmental impacts of buildings provides more confident and correct direction to sustainable architecture. Seriously contemplating on the progress of recent environmental protection around the world, related actions might need further consolidation through more involvement of public sector. To date, the government of Taiwan has initiated several relevant programs, and undoubtedly will continue enhancing and accelerating the environmental protection work based on our hard-won achievements of Green Building development. Furthermore, some of the Green Building indicators have been included in Taiwan's Building Code in 2005. The standard manual for EEWH system was established and popularized widely to building designers, teachers, architects, and contractors. Meanwhile, all the new governmental buildings are required by mandatory Green Building design regulations, and private sector buildings are encouraged to pursue the Green Building Label. Based on these promotion programs, many green remodelling and improvement projects for existing public buildings and schools are currently in progress, and significant savings on electricity and water resources have been achieved. Cooperating with the government, the builders, architects, and even building material manufacturers in Taiwan are working together to reach a better living environment. The Green Building policy is undoubtedly a milestone on the way to upgrading the traditional building industry. The experiences of Taiwan described above can be a reference for other countries in the policy-making of Green Building Development for public sector, especially for tropical/subtropical countries.

REFERENCES

Lin, H.-T. and Yo, M. (1985) A Simplified Seasonal Heat Load Index and its Application to Evaluation of a Building Shelter's Design Condition on a Global Scale, Part 1 & Part 2. *Transaction of Air Conditioning and Sanitary Society of Japan* 59, pp. 47–69.

Lin, H.-T. and Yang, K.-H. (1987) A Simplified Method for Energy Consumption Estimation for Buildings in Hot and Humid

Climates, in *Far East Conference on Air Conditioning in Hot Climates*. Singapore: ASHRAE.

Lin, H.-T., Hsiao, C.-P. and Chen, J.-L. (eds) (2000) The Evaluation System of Green Building in Taiwan in *Sustainable Building 2000 International Conference*, Amsterdam, The Netherlands.

Lin, H.-T. (2004) *Green Architecture in Hot-humid Climates (in Mandarin)*. Taipei: CHAN'S Publishing.

Lin, H.-T. (2005) *Evaluation Manual for Green Buildings in Taiwan (2005 New Edition) (in Mandarin)*. Taipei: Ministry of the Interior.

James, S. (1977) *Sustainable Architecture*. New York: McGraw Hill.

8 IN SEARCH OF A HABITABLE URBAN SPACE-BUILT RATIO: A CASE STUDY OF BUILDING AND PLANNING REGULATION IN DHAKA CITY

Q.M. Mahtab-uz-Zaman[†], Fuad H. Mallick[†], A.Q.M. Abdullah[†]
and Jalal Ahmad[‡]

[†] *Department of Architecture, BRAC University, Bangladesh*
[‡] *Institute of Architects, Bangladesh*

Abstract

Building and planning regulations in Dhaka are weak instruments for designing and managing urban space and built form. Ambiguities in the existing regulations result in the scarcity of open space mainly in residential areas since residential development occupies majority of urban space in order to cater for the growing urban population. The tropical climate suggests the need for sufficient open space with the building forms, to bring about a habitable environment. This chapter highlights the limitation of the existing building regulations, which allow less open space in Dhaka. Case study-based comparison between previous building practices and current standards reveal that certain building regulations have the potential to be revised for better living environment. Some proposals are made emphasizing the need for introducing FAR (Floor Area Ratio); and simulated to suggest a change in the contemporary building practice in order to (a) achieve a comfortable indoor and outdoor environment; (b) create more green areas to reduce urban heat island effect; (c) create a better ecological balance; and (d) preserve low lying areas for water retention as flood protection mechanism – all of which collectively have potentials to enhance social and environmental qualities of the city.

Keywords

Planning and building regulations, tropical climate, built-open area ratio, Floor Area Ratio (FAR), sustainable built form, urban heat island, ecological balance, water retention strategy, indoor environmental quality, social quality.

8.1   INTRODUCTION

This chapter initiates the understanding that there is a disjointed approach to the building regulations that controls the urban built form and open space in Dhaka, the capital of Bangladesh. Results of the inefficiencies are the lack of appropriate social and environmental qualities in urban development practice especially in the context of the tropical climate. Building design seldom follows the design criteria for tropical climates that are necessary to bring a sustainable residential environment. This chapter introduces the gradual development of Dhaka, which experienced a massive urbanization process to respond to the growing population and in-migration from other urban, sub-urban, and rural areas.

8.1.1   Dhaka: Its transformation

As Dhaka has grown through a series of urban expansions coupled with the rapid urbanization and growing population, open spaces have been a neglected issue leading to a situation where built form dominates all parts of the city (Mahtab-uz-Zaman and Lau, 2000a).

Figure 8.1 shows the urban expansion in the city. Further study of the city shows a clear picture of the urban encroachment (Figure 8.2).

8.1.2   Result of transformation

8.1.2.1   *Conservation and heritage*

Very few of the remaining heritage buildings and sites, including the colonial institutional buildings, are fortunate enough to find footprint unharmed. There is very little effort from governmental

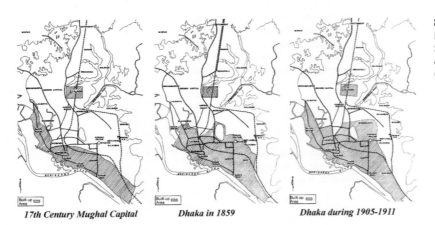

17th Century Mughal Capital   *Dhaka in 1859*   *Dhaka during 1905–1911*

8.1.
Expansion of Dhaka during the Seventeenth to twentieth centuries. (*Source:* Shankland Cox Partnership, 1981)

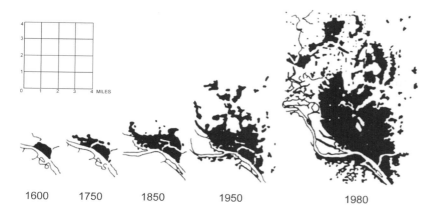

8.2.
Growth pattern of Dhaka during 1600–1980. (*Source:* Shankland Cox Partnership, 1981)

8.3.
Heritage site under pressure by unplanned growth. (*Photo:* Sharif, 2004)

agencies for continuous restoration works. This is due to the fact that the government has more serious urban issues to deal with, in particular, towards the housing of the growing number of people because of urban expansion. Basically, the old part of the city that had more chance for redevelopment could preserve the heritage sites and buildings, unharmed, despite some forms of negligence and vandalism (Figure 8.3). The policies of urban development and expansion never took into account the conservation and heritage issue and will remain so in future.

8.1.2.2   *Urban development and street life*

Besides the negligence towards conservation and heritage, urban development created the opportunity for more inward migration

8.4.
Urban development pressure on street life. (*Source:* Mahtab-uz-Zaman, 2004)

importing many short-lived and mobile trades and businesses. This is evident in street hawking making street life chaotic and unmanageable. Although, New Urbanism (NU) concept (Katz, 1993) supports the basic premise of street life hinged around small retail and hawking, in Dhaka this happened negatively, and in a bigger scale, compounding the problem of urban congestion visibly by obstacles to smooth pedestrian movement (Figure 8.4). This is a common problem in many Asian mega-cities (Mahtab-uz-Zaman, 2000) and requires appropriate regulatory mechanism. (Mahtab-uz-Zaman, 2003a).

8.1.2.3 *Environmental issues*

Immediate surrounding environmental elements, such as, water bodies, trees, green areas could not attract the government's attention; and obviously these became less important issues (Mahtab-uz-Zaman, 1999, 2005a). During the last 5 years, an independent environmentalist group named "Bangladesh Poribesh Andolon (BAPA)" has instigated many movements against filling of water bodies, tree felling, high-density built environments, undesirable construction work at Louis I Kahn's National Assembly Complex, and so forth. This has been a major drive for many professionals to unite and create concern over environmental issues although little has been taken into consideration by the proper authority at the governmental level. Environment is a low priority area for the city developers as this does not bring direct return on their investments.

8.11.
Threat to Nature. (*Source of Figures 8.8–8.11:* A.Q.M. Abdullah, 2004)

8.12.
Satellite photo of the area of studies, showing gradual transformation towards low-rise high-density urban pattern. (*Source:* Abdullah, 2003 and Centre for Environmental and Geographic Information Services – CEGIS, Bangladesh, March 2001)

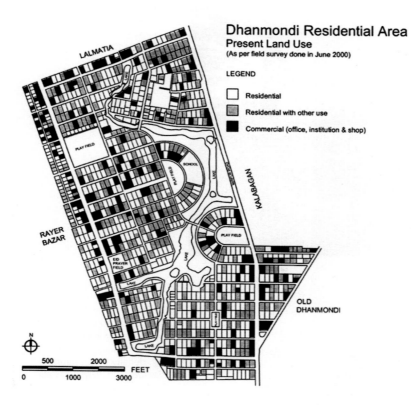

8.13.
Land consolidation by inducing high density built form and reducing open/green space. (*Source:* Hashem, 2001)

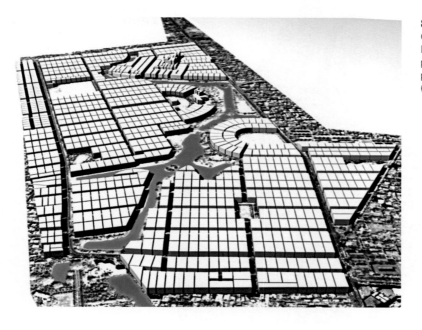

8.14.
Computer simulation of Dhanmondi in 2015 where all plots are developed under the present building by-laws. (*Source:* Abdullah, A.Q.M., 2005)

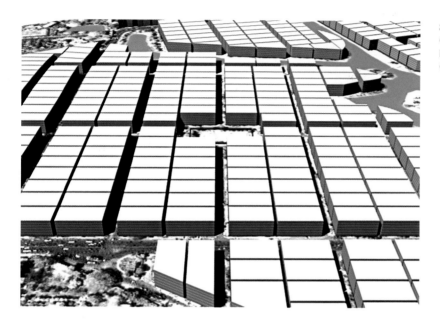

8.15.
Computer simulation, a close-up look of plot-based development (*Source:* Abdullah, A.Q.M., 2005)

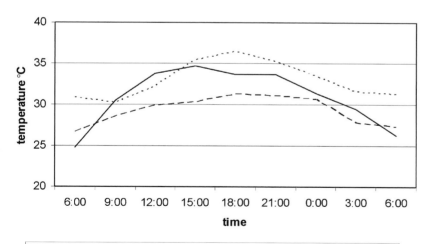

8.16.
Comparison of indoor temperatures in a house in dense surroundings (Figure 8.17a,b) and open surroundings (Figure 8.18) in April. (*Source:* Mallick, 1994)

(a)

(b)

8.17.
House in dense urban setting (1st floor readings). (*Source*: (a) Mallick, 1994, (b) Abdullah, A.Q.M., 2004)

dense urban conditions where buildings are located very close to each other this is not an easy situation to achieve. Koenigsberger (1975) and others in the *Manual of Tropical Housing and Building Part I: Climatic Design suggests* a spacing of 6 times the height between buildings to ensure adequate air flow inside them. Muktadir (1975), based on the results of some wind tunnel tests suggests spacing twice the height between buildings for the same purpose but with particular reference to Dhaka. Research on thermal comfort in urban housing in Dhaka (Figure 8.16) has concluded that in houses or flats in buildings where there are adequate open spaces around, are more comfortable (Figure 8.18) than houses where the buildings are closely together (Figure 8.17(a) and (b)) based on existing set back rules (Mallick, 1994).

8.4 STRATEGIES: ENHANCING ENVIRONMENTAL QUALITIES

8.4.1 Active method

Implementing FAR has proved to be beneficial in many urban development scenarios in various countries (Culpin, 1983; Untermann and Small, 1977; Lynch and Hack, 1984).

Floor Area Ratio (FAR) is defined as:

$$\frac{\text{Gross Square Footage of all structures on a site}}{\text{Gross Square Footage of the lot}}$$

8.18.
Flat in open urban settings (top floor flat readings). (*Source:* Mallick, 1994)

A floor area ratio (FAR) control is a planning tool used to regulate a building's mass in relation to the size of its lot. The FAR is the ratio of the total building floor area to the total lot area. As a result of FAR, the bulk of the structure is reduced in relation to its lot size (Figures 8.19 and 8.20). Communities limit FAR in residential areas in order to avert the proliferation of "gigantic homes" and to ensure a level of development that is compatible with the bulk of existing buildings, and thus create a degree of "visual relief."

8.19.
Concept of FAR. (*Source:* Abdullah, A.Q.M., 2005)

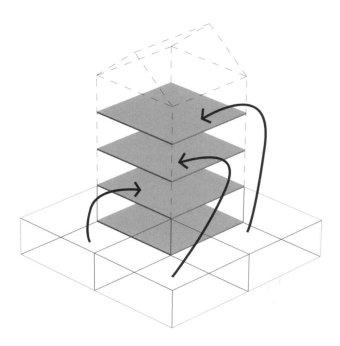

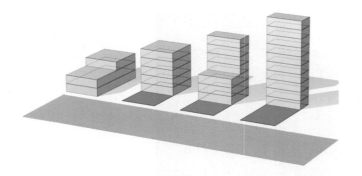

8.20.
Variations in building footprint on a site having a same FAR.
(*Source:* Abdullah, A.Q.M., 2005)

In practice, there is no fixed generalized FAR for different parts of a city. In other words, heights of surrounding properties should determine the maximum allowable FAR for a specific property on a case-by-case basis. Rather, the cities examined set maximum FARs primarily by zoning districts, sometimes in conjunction with neighborhood design guidelines and individual design review. In some cities, the FARs and heights of the surrounding properties are considered in the event that the user wishes to build beyond an allowable FAR (established by zoning districts and neighborhood design guidelines).

In the context of Dhaka, the benefits are no doubt achievable by following the outcomes of FAR method:

- Create air circulation passage/tunnel by increasing set back.
- Create direct sun light passage on open areas by reducing building footprint.
- Reduce urban heat island by re-creating greeneries on open areas.
- Reduce cost of artificial air cooling method as a result of increased natural ventilation.
- Maintain the existing character of established residential neighbourhoods.
- Minimise the out-of-scale appearance of large homes relative to their lot size and to other homes in a neighbourhood.
- Minimise loss of light and privacy to neighbours caused by the construction of large homes.
- Minimise the environmental damage of tree removal and grading or destruction of natural features which may result from overbuilding.
- Permit reasonable expansion of existing dwellings in future.

8.4.2 Passive method

- Reducing building approval process and time by ensuring RAJUK to provide a one-stop service.

- Compliance with the revised building rules.
- Introduction of an Independent Commission for Scrutiny.
- Making RAJUK responsible for non-compliance of the approved plan and ensure professionalism in the design and construction process.
- Introduction of mandatory involvement of enlisted technical personnel and multi-professionals involved in building design, construction and management.
- Introduction of phased application process to ensure appropriateness of development.
- Introduction of Occupancy Certificate (OC) and the mandatory requirement for its renewal every five years to stop unauthorized construction, alteration, and use.

Both Active and Passive methods would collectively influence the urban environment and ensure sustainable urban design qualities by:

- density mix;
- introducing appropriate vehicular circulation and parking requirements for any development;
- allowing architects to explore creative building envelope design and outdoor spaces;
- encouraging road widening and mandatory provision of footpath for easy pedestrian networking; and
- encouraging land pooling/readjustment for a collective housing construction.

But, all the above conditions should be set through introducing incentive packages, such as, allowing more FAR and building height while reducing building footprint.

Figures 8.21–8.29 are computer simulations of different scenarios for similar FAR (more height allow more open space).

8.4.3 Matrix of variations in open area having fixed FAR

Use of FAR requires decision making in terms of how to use residential land effectively. Also, there is a scope of land consolidation in order to achieve a variety of shapes and sizes of open space for environmental and social benefits. For the purpose of demonstration, the authors have used different height limit having different open space combination. For three selected exercises (Figures 8.21; 8.22; and 8.23), it is shown that for the variety of open spaces and built form, a fixed FAR can be maintained. Therefore, the desired FAR target of any builder can be achievable by having different open space-built ratios with the variation in building height. The objective

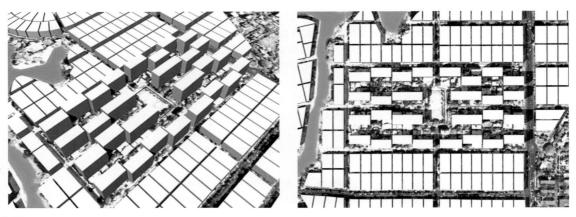

8.21.
Application of FAR (allowing 50% open area with 12-storied height).

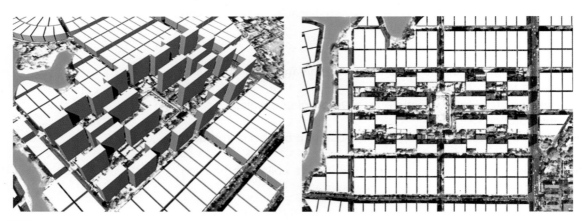

8.22.
Application of FAR (allowing 66% open area with 18-storied height).

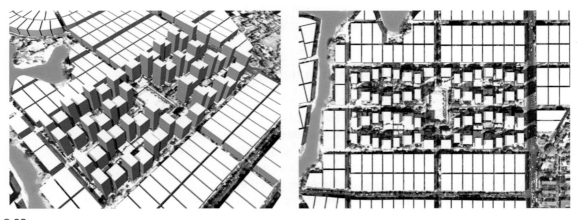

8.23.
Application of FAR (allowing mixed of 50% and 66% open area with 12 and 18-storied height).
(*Source for Figures 8.21–8.23:* Abdullah, A.Q.M., 2005)

of this exercise is to demonstrate that the target investment on land for builders remain unchanged or perhaps, in other words, can be maximized if the designated residential development manage to anchor additional social space by increasing the building height and create room for enhanced environmental qualities.

8.4.4 Enhancing environmental qualities: Scenario

8.5 STRATEGIES: ENHANCING SOCIAL QUALITIES

By introducing FAR in residential development, it is expected to have more open spaces for social activities regardless of size, age, and sex of the residential population.

FAR-generated open space allows more open spaces, water bodies, and pedestrian tracks that generate following activities:

(a) Encourages social activities, such as, children's play area nearby their home, light strolling areas for aged population.
(b) Creates web of jogging tracks along pedestrian way, green and lake areas that allow people to have a healthy atmosphere.
(c) Allows social interaction through face-to-face meeting between neighbours.
(d) Ensures social security by having proper visual and physical linkages through various sizes of connected open spaces.
(e) More open spaces means more green elements [trees and shrubs] that create a balance in nature which, in turn, creates healthy environment for residents.

8.6 CONCLUSION

The underlying difficulty with the building authority in Dhaka or RAJUK is the delay and reluctance of adopting a viable building regulation that complies with the present and future urban development scenario. It is nonetheless significant for RAJUK to embark on postulating the demand for residential building construction required for a population size considered to be optimum for the area under its jurisdiction. Allowing more building height is a means through which the demands of both developers and users are met. It is urgent for RAJUK to amend the building regulation aiming to relax building height and to create opportunity for injecting more open space within the residential areas. New developments in new residential districts which are in their implementation stages need to have a stringent FAR regulation letting the stakeholders know the benefit FAR can offer to the city dwellers (Mahtab-uz-Zaman, 2005).

8.24.
Plot-based development.

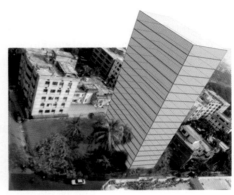

8.25.
FAR-based development.

8.26.
Plot- based development.

8.27.
FAR-based development.

8.28.
Plot-based development.

8.29.
FAR-based development (*Source for Figures 8.24–8.29:* Abdullah, A.Q.M., 2005).

In a broader context, a fragmented institutional network is to be blamed for creating ambiguity in the building regulation framework and implementation process, thereby, creating many obstacles in a viable urban development.

Having such ambiguity in urban development process, the existence and need for a public realm is of less priority for urban managers in the development scheme, result of which is the consuming of open space at a rate that has created a huge urban heat island.

The Dhaka Metropolitan Development Plan of 1995 has no indication of ensuring public realm in urban development. Therefore, there is an urgent need to revise it and to introduce FAR as a key regulation from which other by-laws would generate.

There are examples of good practices in many countries, such as, in Singapore where a proposal has been made for introducing Green Plot Ratio (Ong, 2002). This proposal is appropriate in many ways to enhance the living environment in the tropic and will be able to bring an ecological balance between architecture and urban planning by carefully devising "leaf area index" and "green plot ratio."

The fragmentation and ambiguity in building regulation indicate a less sustainable approach to the city master planning and development. Moreover, formal sector malpractice in distributing plots and green land to influential individuals for development is a regular incident which has massive negative effects on urban development (Khan, 1998).

Although an active real estate market exerts dominance on urban development process, the present land ownership system that has prevailed for several decades create areas of uncertainty and difficulty in implementing viable by-laws. For instance, RAJUK land and privately owned land should have the potential to be pooled together for the larger plots where the implementation of FAR is pragmatic. This makes land readjustment an easier task for the implementation of a planned development. Moreover, environmental and social criteria, which usually have the lowest priority in development, need to be re-addressed by urban managers. For example, the Urban Area Development Plan (DMDP, 1997) recommends adequate measures for preventing the filling-up and reduction of Dhaka's important open spaces and to stop creating new residential plots. Yet RAJUK is under considerable pressure to fill them up, and its shortfall of manpower for development control, and inability to update their plans, has resulted in the lakeside plots being allocated to influential individuals.

A sustainable approach to the desired balance of built-open ratio requires fast implementation of FAR in residential development scheme; and in no way this should be neglected by the concerned urban managers involved in city restructuring process.

ACKNOWLEDGEMENT

The authors acknowledge the support and information provided by the Institute of Architects, Bangladesh and the Research Unit, Department of Architecture of BRAC University.

REFERENCES

Abdullah, A.Q.M. (2003) *Evolution of a Shopping Street – Conflict and Compatibility*, in International Seminar – Architecture Overcoming Constraints, Department of Architecture, Bangladesh University of Engineering and Technology, 11–13 June, Dhaka, Bangladesh.

Ahmad, J., Hossain, Z., Chowdhury, M. (2003) *Building Rules for a Better City Environment*, Daily Star, September, Dhaka, Bangladesh.

Culpin, C. (1983) *Urban Projects Manual*, Liverpool University Press, UK.

DMDP (1997) *Dhaka Metropolitan Development Plan (1995–2015): Urban Area Plan (1995–2015)*, Dhaka Metropolitan Development Planning (DMDP) and Rajdhani Unnayan Kartipakha (RAJUK), Government of the People's Republic of Bangladesh, Bangladesh.

Hashem, M. (2001) *Trends of Development in Dhanmondi Residential Area of Dhaka*, unpublished Master of Urban and Regional Planning Dissertation, Department of Urban and Regional Planning, Bangladesh University of Engineering and Technology, Dhaka, Bangladesh.

Katz, P. (1993) *The New Urbanism: Toward an Architecture of Community*, McGraw-Hill Professional.

Koenigsberger, et al. (1975) *Manual of Tropical Housing and Building. Part 1: Climatic Design*, Orient Longmans, India.

Khan, M.A. (1998) *Rajuk "Secretly" Allotting Plots to High-ups*, Daily Star, 30 April, Dhaka, Bangladesh.

Lynch, K. and Hack, G. (1984) *Site Planning*, 3rd Edition, MIT Press, Cambridge, USA.

Mahtab-uz-Zaman, Q.M. (2005) *Introduce "FAR,"* Letter to Editor, Daily Star, 15 March, Dhaka, Bangladesh.

Mahtab-uz-Zaman, Q.M. (2005a) *Urban Environment – Let's Act Before It's Too Late*, Panorama, The Independent, 4 March, Dhaka, Bangladesh.

Mahtab-uz-Zaman, Q.M. et al. (2004) *In Search for a Habitable Urban Space-Built Ratio: A Case Study of Building and Planning Regulation in Dhaka*, First International Tropical Architecture Conference INTA, 26–28 February, Singapore.

Mahtab-uz-Zaman, Q.M. (2003) *Mass Transit – A Solution to Urban Cholesterol*, Focus, Daily Star, 23 September, Dhaka, Bangladesh.

Mahtab-uz-Zaman, Q.M. (2003a) *Why Cannot We Manage Our City*, Focus, Daily Star, 29 August, Dhaka, Bangladesh. [also available in http://www.dtcb.gov.bd/pollution.htm] and New Age, 13 August, Dhaka, Bangladesh.

Mahtab-uz-Zaman, Q.M. (2003b) *A Note the Roads and Highways Department*, Letter to Editor, Daily Star, 9 June, Dhaka, Bangladesh.

Mahtab-uz-Zaman, Q.M. (2000) Asian Megacities – Reconciliation With the World City Order, *Hinge* 66, pp. 36–52.

Mahtab-uz-Zaman, Q.M. and Lau, S. (2000a) *City Expansion Policy versus Compact City Demand: The Case of Dhaka*. In Compact Cities – Sustainable Urban Forms for Developing Countries, Spon Press, London, edited by Mike Jenks and Rod Burgess, UK.

Mahtab-uz-Zaman, Q.M. (1999) *Accumulated Negligence*, Opinion, Daily Star, 4 April, Dhaka, Bangladesh.

Mahtab-uz-Zaman, Q.M. (1993) *Consolidation as a Response to Urban Growth – A Case in Dhaka*, unpublished Master of Urban Design Dissertation, The University of Hong Kong, Hong Kong.

Mallick, F.H. (1994) *Thermal Comfort for Urban Housing in Bangladesh*, Unpublished doctoral dissertation, Architectural Association Graduate School, London, UK.

Muktadir, M.A. (1975) *Climatic Aspects of High Density Urban Housing in the Warm Humid Tropics with particular reference to Dacca*, Unpublished doctoral thesis, University of Edinburgh, UK.

Ong, B.L. (2002) Green Plot Ratio: An Ecological Measure for Architecture and Urban Planning, *Journal of Landscape and Urban Planning* 63, pp. 197–211.

Shankland Cox Partnership (1981) *Dhaka Metropolitan Area Integrated Urban Development Project*, Report for the Government of Bangladesh, Bangladesh.

Sharif, S. (2004) *Bangladesh from Above*, First Aerial Photo Exhibition, Drik Gallery, Dhaka.

Untermann, R. and Small, R. (1977) Site *Planning for Cluster Housing*, Van Nostrand Reinhold Company, New York.

Part IV

URBAN ENVIRONMENTAL IMPACTS

9 DESIGNING HIGH DENSITY CITIES – PARAMETRIC STUDIES OF URBAN MORPHOLOGIES AND THEIR IMPLIED ENVIRONMENTAL PERFORMANCE

Edward Ng[†], Tak-Yan Chan[†], Vicky Cheng[†], Nyuk-Hien Wong[‡] and Meiqi Han[‡]

[†] Department of Architecture, Chinese University of Hong Kong, Shatin, NT, Hong Kong
[‡] Department of Building, National University of Singapore, Kent Ridge, Singapore

Abstract

Cities of tomorrow must embody the concept of sustainability. Urban design is not about drawing patterns on paper and its architectural studies could not be merely spatial, formal or geometrical. Urban design of the next millennium is about providing and optimising an infrastructure for the enjoyment of its inhabitants while at the same time minimising energy and resources needed, and maximising the benefits of the natural environment. An important consideration of urban design is to provide natural outdoor conditions that are pleasant to human activities. A well-designed outdoor urban environment will also make the design of individual buildings within it easier. Hong Kong and Singapore share common climatic and environmental conditions. Both are cities with limited land resources with a growing and more demanding population. Planning the cities to cope with needs is an important task for their planners. There are many design parameters, for example: Development Density, Plot ratio, Site Coverage, Skyline, Building to Space Ratio, Permeability, Building Shapes and so on. This chapter reports a study based on "skylight" as a design parameter, and how it affects daylight and natural ventilation performances. Experiments were conducted with physical models in wind tunnel and artificial sky, as well as using CFD and computational lighting simulation. The study establishes that, for example, by varying the skylines of the city, the overall daylight and natural ventilation performances could be improved when compared to a city with a uniform skyline. A key message of the chapter is that, through a better understanding, high density cities could be planned and optimised environmentally without losing the development efficacy of the land.

Keywords

Urban design, high density cities, daylight, natural ventilation.

9.1 INTRODUCTION

9.1.1 Background

With the depleting energy and natural resources, and the increasing problem of waste accumulation, cities of tomorrow must embody the concept of sustainability. Urban design is not about drawing patterns on paper and its architectural studies could not be merely spatial, formal or geometrical. Urban design of the next millennium is about providing and optimising an infrastructure for the enjoyment of its inhabitants while at the same time minimising the energy and resources needed and maximising the benefits of the natural environment.

Cities allow goods and ideas to be exchanged effectively. With the increase in world population, many cities are facing the problem of planning to meet with the demands of their inhabitants. Enlarging the city's land boundary, building satellite towns, optimising land zone and usage, and constructing taller and closer packed buildings are some of the tactics used.

High density city design has the advantages of efficient land use, possibility of a viable public transport system, as well as close proximity to amenities and services. For example, Hong Kong is the most energy-efficient metropolis in the world. Its per capita energy consumption is about 40% of the UK, a mere 20% of the USA. One of the key reasons for Hong Kong's outstanding figure is its heavy reliance on public transport. In addition, high-density compact living means that its 7.5 million inhabitants do not have to travel far. The problem of these cities is not how less a high density city might consume. The problem is how much waste it produces "per square metre." And more importantly, how its share of "per square metre" of the available natural elements could be enjoyed among more people.

An important consideration of urban design is to provide natural indoor and outdoor conditions that are pleasant and conductive to human activities (Blocken and Carmeliet, 2004). A well-designed outdoor urban environment will also make the design of individual buildings within it easier. For cities with limited land area, as Hong Kong and Singapore (Figure 9.1), the options open to planners and designers are limited. Finding ways to optimise land use and to design for a higher density seem to be the viable options. For example, Hong Kong is a city of 7.5 million inhabitants living on a collection of islands of a total of 1000 square kilometres. Due to its hilly topography, only 25% of the land is usable. The resulting urban density is around 30 000–50 000 person per square kilometre. Take away the areas for roads, rails, utilities, and open spaces, building sites end up with a net development

9.1.
An example of skyline of Hong Kong.

density of around 3000–4000 person per hectare. This results in high rise buildings, of 40–80 storeys, built close together. Designing and regulating the provision of adequate natural light and air is a difficult task.

Land use density can typically be controlled using plot ratio, site coverage, permissible building volume, maximum building height, street width vs. building height ratio, building profile and set back, and so on. In most countries, regulatory controls are prescriptive and are stated and applied with little understanding of their performance implications. For example, in Hong Kong, the permissible plot ratio of a piece of land is artificially set between 5–10. It is not known how the rules were arrived at in the first place. It is not known what environmental parameters and considerations gave rise to a certain plot ratio to be set for a certain site. It is also little known what would result. In short, one is following the rules without logic, reason or rationality. Hence, it is almost impossible to deal with the regulations intelligently in order to design appropriately.

How some of these planning and design parameters affect the environmental performance of buildings is a research question. This chapter reports pilot parametric studies based on building density, street width, and building skylines. As a demonstration, the study investigated the sensitivity and magnitude of these 3 parameters on the environmental performance of daylight and natural ventilation of buildings.

9.1.2 The parametric methodology

Many scholars worldwide have conducted researches on parametric studies that led to simple design guidelines. Many lasting design understanding and guidelines started with the use of these studies. For example, Givoni (1969) studied the use of wing walls for room air ventilation, which has now been adopted by the Hong Kong Government for its new generation of performance based regulation (Figure 9.2). Hawkes (1970) studied the relationship between block spacing and daylight performance which later led to a site planning guide in UK. Baker and Steemers (2000) studied the relationship between window size and thermal-light energy performance. It resulted in the development of the European LT method and Chan et al. (2001, 2003) studied Urban pollution dispersion using CFD models. This has been quoted in the Urban Design Guidelines of Hong Kong.

Parametric approach is used in this study. The advantage of using parametric study in lieu of studies based on realistic circumstances is that issues could be isolated and simplified to reduce noise and error. It is also much easier to design experimentally. The disadvantage is that results obtained could not be directly and readily related to the real problems. In most cases, results could only indicate the "likely" sensitivity of the performance due to a parameter; but it provides a sense of what is going on and this may initially be sufficient.

For this study, to mimic the conditions of an urban neighbourhood, a 5×5 base plate was used. The base plate had 25 buildings on a square array. The three parameters of density, street width, and building skylines were investigated using a number of simplified scenarios. The three parameters and the three variables gave a permutation of 27 scenarios to be studied (Table 9.1). Three examples of the scenario are shown in Figure 9.3.

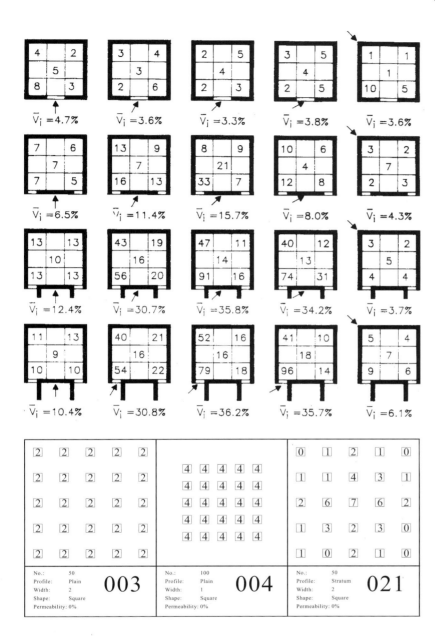

9.2.
Givoni's parametric study of the effect of wing wall and air ventilation of internal spaces.

9.3.
Diagrams show the 3 scenarios tested. For example, scenario 003 represents 25 buildings on a 5 × 5 array on site, each with a height of 2 cubes; all buildings are of equal (uniform) heights; the street to building ratio is 2:1; the blocks are square in shape and there is no gaps/void/hole in the buildings.

For comparative reasons, the scenario of "uniform" height, with "a density of 75 cubes," and with "a street width to building width of 1:1" was regarded the base case (Figure 9.4). This is the kind of urban neighbourhood most likely to have resulted in Hong Kong given the current regulations and control. To mimic the urban surroundings, the base plate was surrounded. For ventilation–wind tunnel study, two additional layers were added. (Figures 9.5 and 9.6) For daylight study, five additional layers were added (Figure 9.7).

Table 9.1. Scenarios of the study 27 scenarios result from a combination of the 3 parameters

Parameters investigated	Variables used		
Density	50 cubes	75 cubes	100 cubes
Skyline	Uniform/Base*	Random**	Stratum***
Building to street ratio	2:1	1:1	1:2

*All buildings are of the same height.
**A random number generator is used to decide the height of each building.
***The heights of the buildings are manipulated manually so that the middle of the site has higher buildings. It resembles a "city centre" condition. The overall effect is still random with varying building heights.

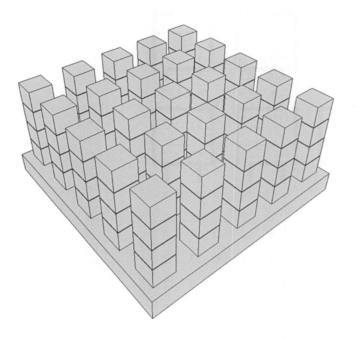

9.4.
The base scenario.

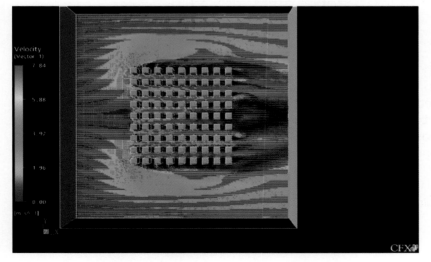

9.5.
An illustration of the base scenario with the surroundings as input during the wind tunnel test. This CFX computational fluid dynamics simulation was also used in the early pilot stage of the study.

9.6.
A scenario (random building height) inside the wind tunnel at NUS.

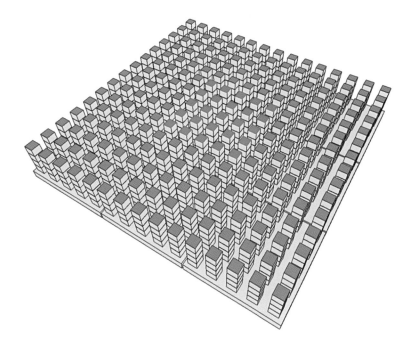

9.7.
A scenario with the surroundings as input during the computational lighting simulation using lightscape.

9.2 DAYLIGHTING

9.2.1 The scientific basis

Daylight performance of an interior space depends on the amount of light available to the vertical surface of the window pane. (Equations 9.1–9.3)

$$E_w = \int_{\theta_L}^{\theta_H} \int_{\phi_L}^{\phi_R} L_{\theta\phi} \cos^2\theta \cos\phi \, d\theta \, d\phi \qquad (9.1)$$

Assuming CIE Overcast Sky:

$$L_{\theta\phi} = L_z \frac{1 + 2\sin\theta}{3} \qquad (9.2)$$

The equation may be resolved to:

$$E = L_z \left[\left\{ \frac{1}{3}(\sin\phi_L + \sin\phi_R) \right. \right.$$
$$\times \left(\frac{\theta_H - \theta_L}{2} + \frac{\sin 2\theta_H - \sin 2\theta_L}{4} - \frac{2\cos^3\theta_H - 2\cos^3\theta_L}{3} \right) \right\}$$
$$+ \left\{ \frac{1}{1 - (0.5\rho_b)} \times \frac{E_{ws}\rho_b}{\pi} \right.$$
$$\left. \left. \times \left(\frac{\pi}{2} - (\sin\phi_L + \sin\phi_R) \times \left(\frac{\theta_H - \theta_L}{2} + \frac{\sin 2\theta_H - \sin 2\theta_L}{4} \right) \right) \right\} \right]$$
$$(9.3)$$

E_w is the illuminance on the window from the sky.
ϕ_H and ϕ_L are the upper and lower angles of obstruction.
θ_R and θ_L are the right and left angles of obstruction.
L_z is the Zentith Luminance.
E_{ws} is the illuminance on the window surface from the sky only.
   This is equal to the the first of the two brackets in Equation (9.3).
$L_{\theta\phi}$ is the Luminance of the patch of sky at θ and ϕ.
ρ_b is the reflectance of the surrounding surfaces.
E is the total illuminance on the window from the sky and reflected light from buildings.

   The amount of light receivable on the vertical surface of the building façade depends on a number of factors. First, the amount of sky the façade "sees," is the Sky Factor (SF). The SF, the amount of sky viewable in terms of solid angle, is corrected using the sky description of the locality to become the Sky Component (SC). Typically the CIE Overcast Sky description is used. (Tregenza, 1999) (Figure 9.8) This description depicts the sky condition of a dull cloudy day without direct sunlight. The amount of light available under this CIE overcast sky is azimuth independent. Many artificial skies and computational programme are built based on this sky type. Second, in dense urban conditions, most available light is the reflected light (ERC) of surrounding surfaces. It depends on the reflectivity of the surfaces, as well as how well these surfaces are illuminated directly in the first place. It has been calibrated in a previous study relating Sky Component, reflectance of surrounding surfaces, and Vertical Daylight Factor (VDF) of the building façade. [Ng, 2003] Hence, knowing SC, VDF could be computed easily. (Figure 9.9)

9.2.2   The studies – skyline

Computational lighting simulation was used to conduct tests of the 27 scenarios. Lightscape had been selected. (Figure 9.10) The software had been calibrated to yield good results under heavily obstructed conditions [Ng, 2001]. The geometry was modelled in FormZ solid modeller and exported .dxf into Lightscape. All surfaces were oriented and assigned a reflectance of 0.2. Rendering with cloudy sky with sun illuminance set to 0, the results were picked using the Analysis Function of the programme. The illuminance of all four surfaces of all the blocks of the 5 × 5 grid on the base plate was recorded. Thus for a study with 75 blocks, 300 data points could be recorded.

9.2.3   Results

Results are shown in Figure 9.11, 9.12 and 9.13 below.

Take Figure 9.11 (uniform height) as an example, it illustrates that light levels generally fall into 3 distinct bands: top, middle, and bottom. This roughly depends entirely on the levels of the blocks. Roughly a third of the data points are in the low ranges in the order of 8–9% Vertical Daylight Factor (VDF). Results of the random, and stratum scenarios (Figure 9.12 and 9.13) illustrate that light performance spread out and distribute along the x-axis. There is no clear pattern relating the level of the blocks and their performances. Summing the light performance of all the surfaces reveals that the median of uniform (base), random, and stratum scenarios (when street:building is 1:1, density is 75 blocks) are 15.2, 16.7 and 17.6% respectively. This means that on the whole light performance of

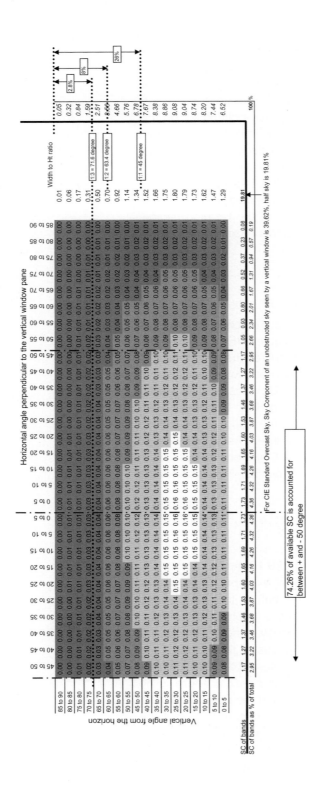

9.8.
The table summarises the amount of light a vertical window can receive under a CIE Overcast Sky.

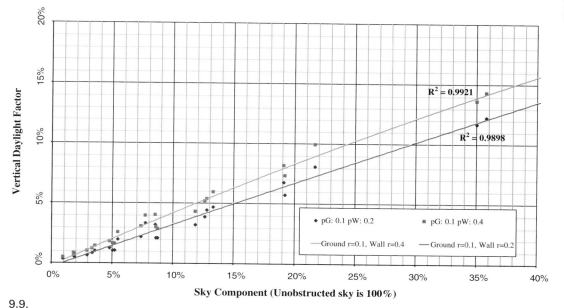

9.9.
Sky Component (*x*-axis) vs. Vertical Daylight Factor (*y*-axis) under 2 different assumptions of reflectance of surrounding buildings (*r* = 0.2 [bottom line], *r* = 0.4 [top line]).

9.10.
One of the 27 scenarios tested.

stratum scenario is roughly 20% better than the uniform (base) case. It signifies that there may be benefits optimising the "height difference" parameter.

Results of the 27 scenarios tested are summarised in Table 9.2. It is noted that in all cases, light performance of stratum scenarios exceed the random scenarios, which in turn exceed the base scenarios. For example, in very high density conditions of $5 \times 5 \times 4$ and a street width to building width ratio of 1:1, the performance of the stratum layout is around 30% better than the base case. That is to say, given the same design density, one design is better than the other by roughly a third. Alternatively, examining the light performance of $5 \times 5 \times 3$ base and $5 \times 5 \times 4$ stratum, the two sets of data are very similar. That is to say, given a certain daylight performance requirements, one could build either 75 blocks all with the same building height, or 100 blocks with varying building heights.

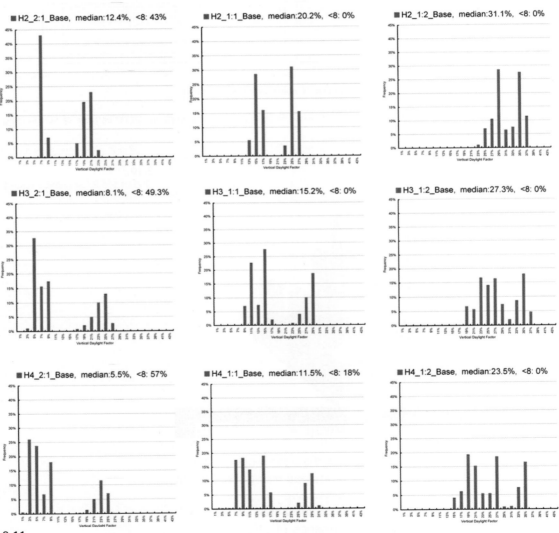

9.11.
Scenarios with uniform heights. VDF performance (*x*-axis) was plotted against the cumulative occurrence (*y*-axis).

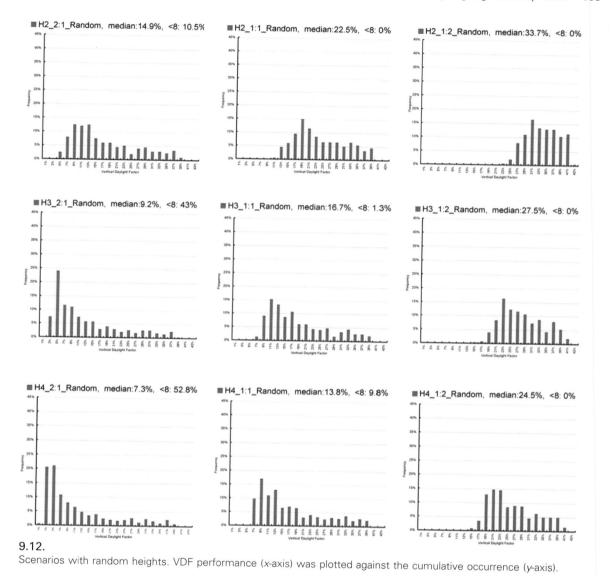

9.12.

Scenarios with random heights. VDF performance (x-axis) was plotted against the cumulative occurrence (y-axis).

When plotting the 27 scenarios along the VDF y-axis of Figure 9.14, it is noted that, in general, light performance shift-up from base, random to stratum (Blue lines). The improvement is roughly 10–30%. Regarding street width (Red lines), when the street widen from 1 to 2, there is a marked and significant improvement in terms of VDF. However, when the street narrow from 1 to 0.5, there is a decrease, but not with the same magnitude. Regarding density (Green lines), there is roughly a 100% difference when doubling density from 50 to 100. Higher density and narrower streets have effects of VDF performance of buildings. By varying

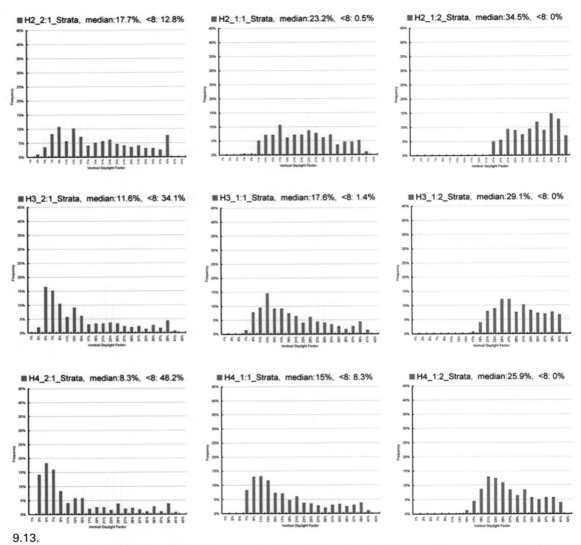

9.13.
Scenarios with stratum heights. VDF performance (x-axis) was plotted against the cumulative occurrence (y-axis).

Table 9.2. Medians of VDF of all 27 scenarios

	Street: Building	Base	Random	Stratum
5 × 5 × 2 blocks high	2:1	31.3	33.7	34.5
	1:1	20.2	22.5	23.2
	1:2	12.4	14.9	12.8
5 × 5 × 3 blocks high	2:1	27.3	27.5	29.1
	1:1	15.2	16.7	17.6
	1:2	8.1	9.2	11.6
5 × 5 × 4 blocks high	2:1	23.5	24.5	25.9
	1:1	11.5	13.8	15
	1:2	5.5	7.3	8.3

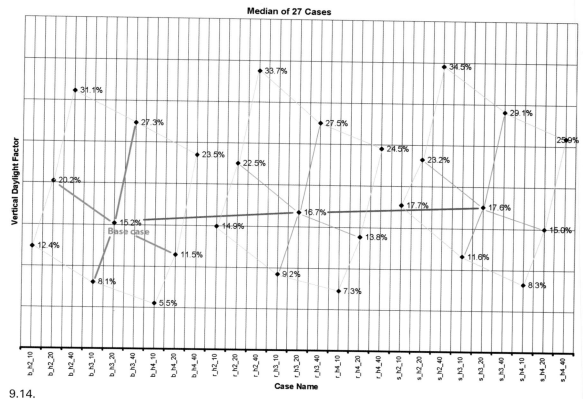

9.14.
Medians of VDF of 27 scenarios. Note the general shift from left to right.

to building heights, one could improve performance. For example, the effect of having a stratum skyline means that one could offset the effects of density increase from 75 to 100.

9.2.4 Sensitivity of building height variation and lighting performance

Noting the potential benefits of varying the skylight, the second question is: how much to vary? A similar experiment to those described above is used. This time the number of cubes is fixed at 100. The street width is 1. A random number generator is used to generate 16 scenarios of different height variations. (Figure 9.15)

9.2.5 Results

In this experiment, VDF readings of the four surfaces of the bottom 25 cubes are taken and tabulated. (Figures 9.16–9.17) It can be noted that the minimum decreases gently when height difference decrease. On the other hand, the maximum decreases more rapidly.

A	B	C	D	E	F	G	H	I	J	K	L	M	N	O	P
4	7	4	7	4	3	4	4	3	4	2	6	5	2	0	3
3	4	3	6	6	3	1	7	3	3	3	7	5	1	0	4
3	2	2	8	5	5	2	3	4	4	5	2	6	3	7	5
4	2	7	3	2	4	1	6	3	6	2	1	3	2	2	4
4	2	1	7	3	3	5	3	6	5	1	6	3	3	0	4
4	2	3	2	7	3	2	3	3	3	1	6	6	8	1	3
5	5	2	3	2	4	7	1	3	3	8	4	3	7	8	5
3	7	2	1	3	5	2	2	4	3	2	2	3	3	5	5
4	4	3	1	3	5	4	1	4	3	8	8	5	7	3	3
3	4	3	2	6	3	2	7	4	6	7	4	4	2	1	5
5	4	2	5	4	3	6	7	6	4	5	5	3	1	9	4
4	5	5	0	2	5	5	6	4	2	2	4	6	5	8	4
3	7	4	7	4	5	5	6	6	3	6	7	3	4	5	4
4	2	2	8	6	5	1	7	3	5	1	3	4	8	2	5
4	5	1	4	5	6	5	4	3	3	5	4	5	4	6	3
4	6	5	3	2	3	4	3	5	2	5	1	2	7	3	5
4	5	6	4	4	3	6	7	4	6	3	3	3	3	3	3
5	4	6	2	2	4	7	5	3	4	7	7	4	7	7	4
5	2	4	4	2	3	2	7	4	5	7	6	3	7	3	3
3	3	7	8	7	3	7	2	5	3	3	0	6	1	9	4
3	4	6	0	3	3	7	0	3	5	3	5	5	2	0	3
4	4	6	4	4	4	2	3	5	5	7	1	2	1	6	5
4	2	7	8	5	5	6	3	4	6	4	0	2	4	1	4
5	5	5	1	7	6	5	3	3	5	2	9	3	2	3	3
4	5	4	2	2	4	2	3	3	2	1	5	6	6	8	4
2	5	6	8	5	3	6	7	3	4	7	9	4	7	9	2

9.15.
Scenarios A to P for the height sensitivity study. Heights of the 25 buildings are shown.
Numbers in Red (bottom) is the height difference, from 2 to 9.

Height difference	Case	Min	Max	Mean	Median	Standard deviation
9	L	8	1 9	14.1	14	1.45
9	O	7	2 2	14.4	14	1.42
8	D	9	1 7	12.1	13	1.27
7	H	7	1 6	11.6	11	1.29
7	K	6	1 7	11.2	11	1.31
7	N	9	1 7	11.9	12	1.19
6	C	8	1 5	11.4	11	1.12
6	G	7	1 4	10.8	11	1.02
5	B	8	1 3	10.5	10	1.07
5	E	7	1 4	11.0	11	0.98
4	J	8	1 3	10.4	10	0.92
4	M	8	12	9.9	10	0.88
3	I	8	1 2	10.0	10	0.91
3	F	7	11	9.2	9	0.87
2	A	6	11	8.7	9	0.82
2	P	6	10	8.3	8	0.80

9.16.
Results of the height sensitivity study.

And overall, there is an almost linear relationship between height difference and VDF performance. The study indicates that in general, the "more difference" the better.

9.2.6 Discussion

The parametric study of skyline and building heights reveals that "height difference" can be a useful design parameter to optimise

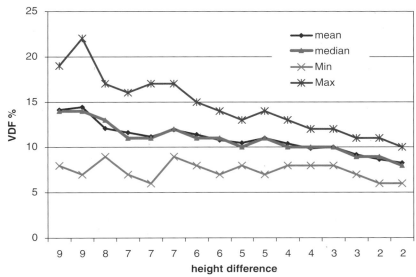

9.17.
VDF trend lines of various height difference scenarios.

an already high density city. In hindsight, this seems obvious. Take the Sky Component Table of a CIE Overcast Sky and reproduce as an illustration (Figure 9.18). For example there are buildings in front of the window. The buildings on the left are volumetrically the same as the buildings on the right – same density. The buildings on

Horizontal angle perpendicular to the vertical window pane

Vertical angle from the horizon	45 to 50	40 to 45	35 to 40	30 to 35	25 to 30	20 to 25	15 to 20	10 to 15	5 to 10	0 to 5	0 to 5	5 to 10	10 to 15	15 to 20	20 to 25	25 to 30	30 to 35	35 to 40	40 to 45	45 to 50
85 to 90	0.00	0.00	0.00	0.00	0.00	0.00	0.00	0.00	0.00	0.00	0.00	0.00	0.00	0.00	0.00	0.00	0.00	0.00	0.00	0.00
80 to 85	0.00	0.00	0.00	0.00	0.00	0.01	0.01	0.01	0.01	0.01	0.01	0.01	0.01	0.01	0.01	0.00	0.00	0.00	0.00	0.00
75 to 80	0.01	0.01	0.01	0.01	0.01	0.01	0.01	0.01	0.01	0.01	0.01	0.01	0.01	0.01	0.01	0.01	0.01	0.01	0.01	0.01
70 to 75	0.02	0.02	0.02	0.02	0.02	0.03	0.03	0.03	0.03	0.03	0.03	0.03	0.03	0.03	0.03	0.02	0.02	0.02	0.02	0.02
65 to 70	0.03	0.03	0.03	0.04	0.04	0.04	0.04	0.04	0.04	0.04	0.04	0.04	0.04	0.04	0.04	0.04	0.04	0.03	0.03	0.03
60 to 65	0.04	0.05	0.05	0.05	0.05	0.06	0.06	0.06	0.06	0.06	0.06	0.06	0.06	0.05	0.05	0.05	0.05	0.05	0.05	0.04
55 to 60	0.05	0.06	0.06	0.07	0.07	0.07	0.08	0.08	0.08	0.08	0.08	0.08	0.08	0.08	0.07	0.07	0.07	0.06	0.06	0.05
50 to 55	0.07	0.07	0.08	0.08	0.09	0.09	0.09	0.10	0.10	0.10	0.10	0.10	0.10	0.09	0.09	0.09	0.08	0.08	0.07	0.07
45 to 50	0.08	0.09	0.09	0.10	0.10	0.11	0.11	0.11	0.12	0.12	0.12	0.12	0.11	0.11	0.11	0.10	0.10	0.09	0.09	0.08
40 to 45	0.09	0.10	0.11	0.11	0.12	0.12	0.13	0.13	0.13	0.13	0.13	0.13	0.13	0.13	0.12	0.12	0.11	0.11	0.10	0.09
35 to 40	0.10	0.11	0.11	0.12	0.13	0.13	0.14	0.14	0.14	0.14	0.14	0.14	0.14	0.14	0.13	0.13	0.12	0.11	0.11	0.10
30 to 35	0.10	0.11	0.12	0.13	0.14	0.14	0.15	0.15	0.15	0.15	0.15	0.15	0.15	0.15	0.14	0.14	0.13	0.12	0.11	0.10
25 to 30	0.11	0.12	0.12	0.13	0.14	0.15	0.15	0.15	0.16	0.16	0.16	0.16	0.15	0.15	0.15	0.14	0.13	0.12	0.12	0.11
20 to 25	0.11	0.12	0.12	0.13	0.14	0.14	0.15	0.15	0.15	0.16	0.16	0.15	0.15	0.15	0.14	0.14	0.13	0.12	0.12	0.11
15 to 20	0.10	0.11	0.12	0.13	0.13	0.14	0.14	0.15	0.15	0.15	0.15	0.15	0.15	0.14	0.14	0.13	0.13	0.12	0.11	0.10
10 to 15	0.10	0.10	0.11	0.12	0.13	0.13	0.14	0.14	0.14	0.14	0.14	0.14	0.14	0.14	0.13	0.13	0.12	0.11	0.10	0.10
5 to 10	0.09	0.09	0.10	0.11	0.11	0.12	0.12	0.13	0.13	0.13	0.13	0.13	0.13	0.12	0.12	0.11	0.11	0.10	0.09	0.09
0 to 5	0.08	0.08	0.09	0.09	0.10	0.10	0.11	0.11	0.11	0.11	0.11	0.11	0.11	0.11	0.10	0.10	0.09	0.09	0.08	0.08

9.18.
Benefits of varying building heights in high density cities.

the left hand side are of equal heights and externally obstruct the window to 60°. The buildings on the right are of different heights. Some are very tall, and others are very low. It can be illustrated that the VDF performance on the right will be much better – as the tall buildings block lesser and lesser light as it goes up while more and more light could be gained with a decrease in height of its adjacent building.

9.3 NATURAL VENTILATION

9.3.1 The studies – skyline

Wind tunnel has been used for ventilation studies. 27 scenarios of 3 parameters (Density, Skylight and Building to Street ratio) are conducted (Table 9.3). Similar to the daylight study, the 75 cube, uniform and 1:1 scenario is used as the base case.

While Vertical Daylight Factor (VDF) has been used as a criterion to assess daylight performance, Air Change per Hour (ACH) is used as a criterion for wind and ventilation studies (Awbi, 1991). ACH is calculated using Contam96. ACH, which is the reciprocal of age of air (ASHRAE Handbook, 1997), determines the ventilation performance of an indoor space. It is a convenient indicator for the study.

Simulation of ACH values using Contam96 requires the input of Wind pressure (P_w), which can be calculated by using equation (9.4). C_p values can empirically be obtained either by field measurements or wind tunnel tests. In this study, C_p values are obtained via wind tunnel test as described below.

$$P_w = (\rho \times C_p \times v^2)/2 \qquad (9.4)$$

P_w: wind pressure (Pa)
C_p: pressure coefficient
ρ: density of air (= 1.20 kg/m^3)
v: wind speed at reference point in wind tunnel (m/s)

Table 9.3. Scenarios of the study

Parameters investigated	Variables used		
Density	75	100	125
Skyline	Uniform	Random	Stratum
Building to street ratio	2:1	1:1	1:2

Wind tunnel is used as it is a reliable tool for the study (ASCE, 1982; Plate, 1999). A physical scale model of the test area and its surroundings were constructed and placed in an open circuit boundary layer wind tunnel at the National University of Singapore. The wind tunnel dimension is 17 m length × 3.75 m width × 1.75 m height (Figures 9.19 and 9.20). The devices listed in Figure 9.20 are necessary to simulate the atmospheric boundary layer, close to the ground, at a prescribed turbulence level and length scale at various points in the test section. They must be carefully calibrated before the test is conducted. As shown in Figure 9.21, the scale model was subjected to a controlled wind flow. The pressure reading at each pre-installed sensor tap, corresponding to the ventilation opening on

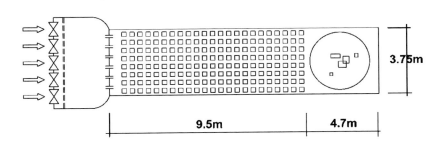

9.19.
Plan of the wind tunnel.

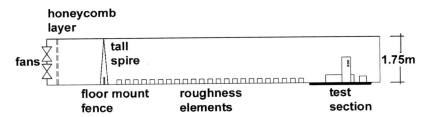

9.20.
Section of the wind tunnel.

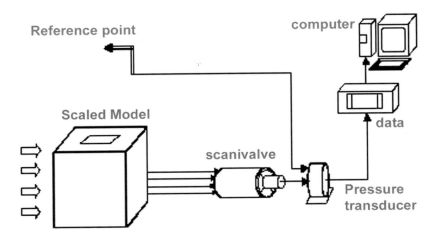

9.21.
Equipment set up.

building envelop, was recorded by the scanivalve with a sample size of 25. This was then converted to electrical signal by the pressure transducer. Through the data acquisition system, electrical signal were acquired and converted into C_p.

Specifying the reference level for wind velocity is essential for wind pressure calculation, since the surface roughness and height above the ground can influence the strength of wind (BS 5925, 1991). The wind velocity (Uz) at the reference height in the wind tunnel is computed with equation (9.5). Equation (9.5) provides an approximation to account for height difference and intervening terrain between "on-site" wind data and reference point in wind tunnel (BS 5925, 1991).

$$U_z = U_m \times k \times z^a \qquad (9.5)$$

U_m: 2.5 m/s (assume average wind speed at a standard height of 10 m)
z: 90 m (height of reference point in wind tunnel)
k, a: 0.35 and 0.25, respectively in the urban context (constants dependent on terrain)

The layout of typical level (Figure 9.22) was required for Contam96 simulation. The building was modelled as a number of zones depend on the layout of the building and zoning of the ventilation system. Assuming each level was divided into 4 equally-spaced unit, the mean C_p value at each unit was converted to P_w and Contam96 was used to simulate the age of air in each zone by using P_w obtained from the external envelope of building. The same P_w was used for the two fenestrations on the particular façade. The overall mean ACH (Z1–Z4), the windward mean ACH (Z1 and Z2) and the leeward

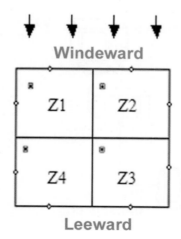

9.22.
Typical level layout for Contam96.

mean ACH (Z3–Z4) were calculated for all the cases. They represent the ventilation performance of buildings. The following properties were input into the indoor zone, airflow path and airflow element:

Zone data	• Temperature at 23°C
	• Variable pressure
	• Volume of 147 m^3 or floor area of 49 m^2
	• Variable contaminant concentrations
Airflow path	• Relative elevation of 1 m
	• Multiplier of 1
	• Constant wind pressure (Pa) – input of P_w data
Airflow element	• 1.8 m^2 cross sectional area (fenestration is assumed to be 1.5 × 1.2 m)
	• 0.5 flow exponent
	• 0.6 discharge coefficient
	• Transition Reynolds number of 30
	• Hydraulic diameter of 1.34164 m

9.3.2   Results

Results of four scenarios are shown in Figures 9.23–9.25 as examples. It can be seen that most of the points in the uniform base case have air change rates in the low ranges whereas random and stratum cases show a more evenly distributed ventilation performance. When compared to stratum case, random case shows less frequency of occurrence towards the lower range and more frequency towards the higher range for all the street widths.

Table 9.4 shows the average of the ACH of all the units. Random case and stratum case have similar air change rate values. However, it can be seen that the random case has the highest value and hence

Table 9.4. Average of Air change rate of all 27 scenarios

	Street: Building	*Base*	*Random*	*Stratum*
5 × 5 × 5 cubes	0.5:1	11.1	17.3	16.8
	1:1	12.3	18.9	17.6
	2:1	17.4	**24.8**	22.3
5 × 5 × 4 cubes	0.5:1	11.0	18.7	18.3
	1:1	13	19.5	19
	2:1	16.8	**23.3**	22.7
5 × 5 × 3 cubes	0.5:1	11.0	18.5	17.6
	1:1	14.3	19.3	19.3
	2:1	15.6	**22.8**	21.4

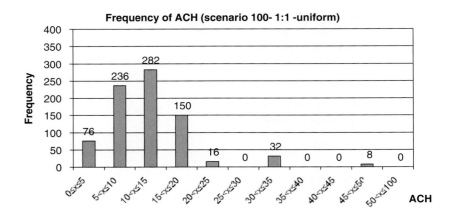

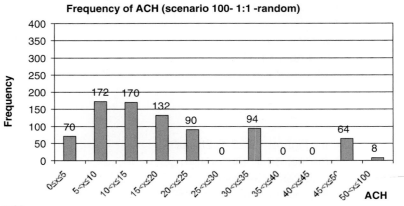

9.23.
Air change rate (x-axis) against the cumulative occurrence (y-axis). Comparing uniform (plain) height with random heights.

the best ventilation performance for all the street width. For street to building width ratio of 1:1, density of 100, the average ACH is 13, 19.5 and 19 for base, random and stratum cases, which shows that random case is 50% and stratum case is 46% better than base in terms of ventilation performance. Comparing the results of all the street width to building ratio, as expected, the case with 2:1 has the maximum ACH value and 0.5:1 has the minimum value for base, random and stratum cases. Among all the cases, the best ventilation performance was achieved with random heights and 2:1 street to building width ratio. For density of 100, random is 79% better than the base case. For the street to building width ratio of 0.5:1, ACH increased by 70% and 66%, for random and stratum cases respectively.

Refer to Figure 9.26, the solid (blue) trend lines indicate the improvement due to building height variation changes. It is

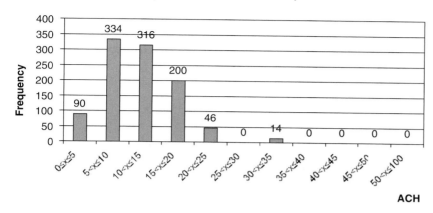

Frequency of ACH (scenario 125 - 1:1 - uniform)

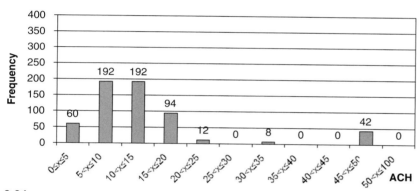

Frequency of ACH (scenario 75 - 1:1 - uniform)

9.24.
Air change rate (x-axis) against the cumulative occurrence (y-axis). Comparing density of 75 cubes with 125 blocks.

significant. The dash lines (green lines) show the effects of street width changes. This is the next significant parameter. The least important parameter seems to be the density change (red lines). For example (as shown at the bottom left corner data points), at narrow street to building width of 0.5:1, given uniform building heights, density change has no effect to overall average ACH.

It can easily be summarised that significant overall average ACH improvement could be achieved by varying the building heights. Improvements of some 70% or more is remarkable when compared, with the earlier VDF study.

Further data analysis is conducted. A question is asked if improvement is achieved where it is needed most. in other words, are the units on the lower floor receiving the benefits? Table 9.5 reveals that the improvement from base to random and stratum is only around

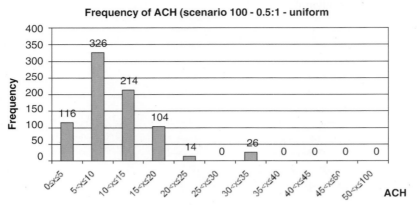

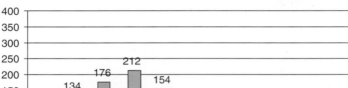

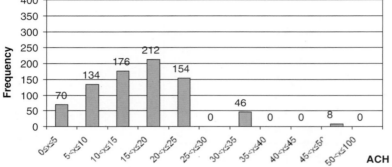

9.25.
Air change rate (*x*-axis) against the cumulative occurrence (*y*-axis). Comparing street to building width of 2:1 and 0.5:1.

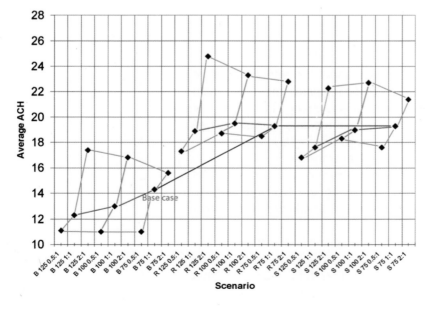

9.26.
Overall average Air change rates (ACH) of all 27 scenarios.

Table 9.5. Average of Air change rate of all 27 scenarios – bottom 2 levels

	Street: Building	Base	Random	Stratum
5 × 5 × 5 cubes	0.5:1	11.0	13.9	13.2
	1:1	12.9	13.9	12.9
	2:1	17.1	18.1	17.6
5 × 5 × 4 cubes	0.5:1	12.5	14.5	14.7
	1:1	14.7	13.8	14.4
	2:1	17.2	17.5	17.6
5 × 5 × 3 cubes	0.5:1	11.4	13.4	13.4
	1:1	15.4	13.6	14.7
	2:1	15.9	17.1	16.5

5–10%. Very much in line with the observation of the daylight VDF study, the minimum will stay roughly the same. The improvement is captured by the increase of the maximum – thus pushing up the median and the average.

From Figure 9.27, it is immediately noticeable that the average ACH of lower levels are lower, especially for the random and stratum scenarios. However, ACH of the base scenario of average and "lower levels" are similar. Great improvements could be attained by increasing the street width. Once the street width is narrower at 1:1 or lower, the effect stabilises. The most important to note is that varying heights of buildings has relatively little

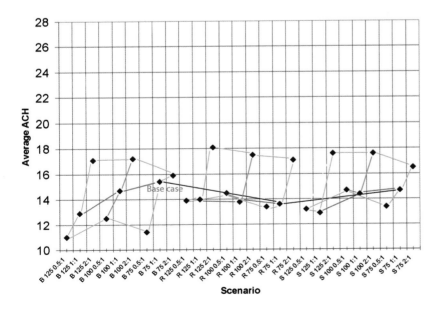

9.27.
Overall average air change rates (ACH) (bottom two levels only) of all 27 scenarios.

effect. Some improvements are noted for scenarios of very narrow streets (0.5:1).

The results of the lower levels contrast greatly with that of the overall average. It may be concluded that varying the building heights could improve the overall ACH of the city, however, the benefits are not felt where it is needed most – lower levels; except perhaps when the streets are very narrow compared with the building heights.

9.3.3 Sensitivity of building height variation and ventilation performance

Similar to the daylight study, the question is asked: how does the ACH performance relates to the height differences of the city? Is there an optimum height? The following experiments are conducted.

Capturing ACH of all cubes in the city is experimentally time-consuming. The equipment at the NUS wind tunnel could only monitor 48 channels in one way. Shifting the pressure valves is tedious. In order to simplify the experimental procedure, a random selection of cubes will be monitored. The pressure valves will be fitted only to some cubes of the 2 lower levels. Z tests indicate that this "reduced" sample set of readings with 46 data points could represent the results of the population of various densities.

Nine scenarios are tested, all with a street:building width of 1:1. They all have a density of 100 cubes. The uniform/base scenario has a height contrast of 0, whereas on the other extreme, a contrast of 14 is tested. Table 9.6 summarises the results.

Table 9.6. ACH performance of lower levels with different height contrasts

Height contrast	Height difference Max:min	ACH
0	4:4	10.5
3	3:6	10.8
4	3:7	11.9
6	2:8	13.8
7	2:9	11.2
8	1:9	13.3
10	1:11	13.4
10	0:10	17.9
14	0:14	17.0

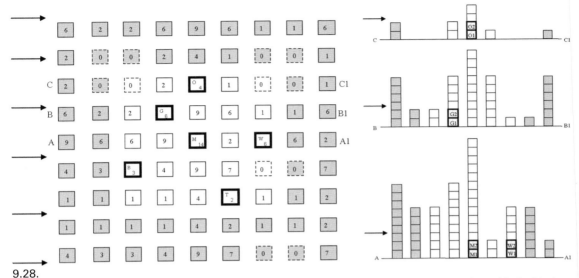

9.28.
Scenario 0:10. The grey cubes are surroundings. The 25 white cubes represent the city. The cubes with the black border are where the sensors are located. Dotted line cubes are open space. The number in the cube indicates the height of the cube. The 3 sections on the right cut across the site. For example C–C reveals the open spaces of the city, as well as the surroundings.

There is a general improvement at the lower levels when the height contrast increases. The improvement is upward of 30%, and could go up to 70%. Between the height differences of 4:4 to 1:11, ACH performance fluctuates among 10–13. The data is a bit confusing and seems to tally with the earlier experiments of 27 scenarios. However, a great jump of ACH from 13 to 17 is noted with the height difference of 0:10 and 0:14.

The most important observation is the two scenarios of height contrast of 10 (1:11 vs. 0:10). It is highly likely that the results are not due purely to the height contrast. It is speculated that it is the "zero" that might have caused the differences. "zero" in this case means an open space in the city. For example, the 0:10 scenario has 3 open spaces whereas the 1:11 has none (Figure 9.28).

9.4 CONCLUSION

9.4.1 Discussions

The study demonstrates beyond reasonable doubt that for lighting and air ventilation, better overall performance could be obtained by varying the skylines. The improvement is around 20–30% for daylight, and 35–70% for air ventilation. Thus, it is important to

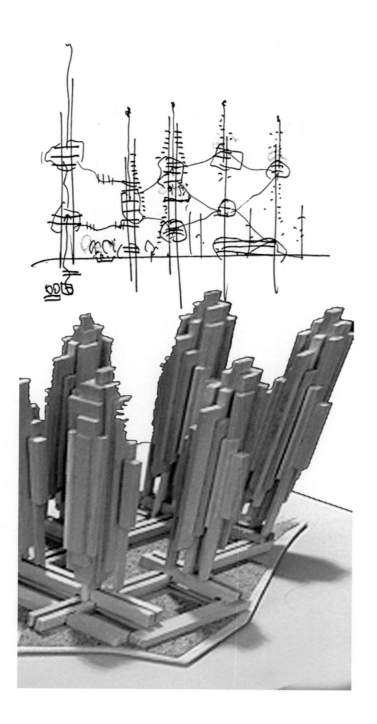

9.29.
A sketch of what a high density city optimised for light and air could look like. The unique morphology will have to be further understood architecturally.

capitalise this by effecting design guides, building and planning regulations to encourage that to happen. The study also reveals that for lower levels, the improvement for ventilation could be marginal and the results indicate that there may be other more important parameters.

The results so far are preliminary, but the study illustrates an approach to designing high density city. With limited resources, optimisation is the key word. Cities in the past could be designed with a lot of redundancy. Most of the time, this is not an important issue, for the cost benefit considerations are not critical. However, like racing a Formula 1 car, when every millisecond counts, design will have to take on a completely new agenda (Figure 9.29).

9.4.2 Further works

This study only identifies preliminarily the quantitative effects of three parameters. Further studies are needed to include additional scenarios and to further investigate the precise mathematical relationship between the parameters and performances. For example, if building skyline is an important parameter as stated in this study, what is the range of building heights one should recommend, and if there are simple mathematical formulae to describe the relationship? Can an optimum solution be found and expressed mathematically? This parametric study is a beginning. It investigated density, building skylines and street width to building width ratios. Other parameters like building shapes, gaps between buildings, permeability of the urban fabric and building surfaces could also affect daylight and air ventilation performances. They require further studies.

ACKNOWLEDGEMENT

Thanks are due to Vicky Cheng, Meiqi Han, Tak-Yan Chan assisting the research. The study was funded by a Direct Grant of CUHK.

REFERENCES

ASCE (1982) Manuals and Reports on Engineering Practice No. 67, Wind Tunnel Studies of Buildings and Structures. American Society of Civil Engineering.

American Society of Heating, Refrigerating and Air-Conditioning Engineers Handbook – Fundamental (Loads and energy calculation) (1997), Chapter F25 Ventilation and infiltration.

Awbi, H.B. (1991) *Ventilation of Buildings* (1st ed.). London: Spon Press.

Baker, N. and Steemers, K. (2000) Energy and Environment in Architecture – a Technical Design Guide, London: E & FN Spon.

Blocken, B. and Carmeliet, J. (2004) Pedestrian Wind Environment Around Buildings: Literature Review and Practical Examples. *Journal of Thermal Envelope and Building Science* 28, no. 2, pp. 107–159.

British Standard Institution (1991) Code of Practice: Ventilation Principles and Designing for Ventilation BS 5925. British Standard U.K.

Chan, A.T., So, E.S.P. and Samad, S.C. (2001) Strategic Guidelines for Street Canyon Geometry to Achieve Sustainable Street Air Quality. *Atmospheric Environment* 35, pp. 5681–5691.

Chan, A.T, Au, W.T. and So, E.S.P. (2003) Strategic Guidelines for Street Canyon Geometry to Achieve Sustainable Street Air Quality – Part II: Multiple Canopies and Canyons. *Atmospheric Environment* 37, pp. 2761–2772.

Givoni, B. (1969) *Man, Climate and Architecture*. Elsevier.

Hawles, D. (1970) Factors Affecting the Bulk and Separation of Buildings, PhD Dissertation, Cambridge University, Cambridge, U.K.

Ng, E. (2001) Daylighting Simulation of Heavily Obstructed Residential Buildings in Hong Kong in Vol. 2, Lamberts, R., Negrao, C.O.R., Henson, J. (eds), *Proceedings of International Building Performance Simulation Association Conference*, Rio, Brazil, pp. 1215–1222.

Ng, E. and Chan, T.Y. (2003) A Simple Method for Estimating Daylight Availability, in *Proceedings of New Technology for Better Built Environment*, Mainland-Hong Kong Joint Symposium, HKIE-BSD, ASHRAE-HKC, CIBSE-HKB, Shandong HVAC, Qingdao, China, pp. A82–A91.

Plate, E.J. (1999) Methods of Investigating Urban Wind Fields – Physical Models, *Atmospheric Environment* 33, pp. 3981–3989.

Tregenza, P. (1999) Standard Skies for Maritime Climates, LRT 31, no.4.

Wong, N.H., Feriadi, H., Tham, K.W., Sekhar, S.C., Cheong, K.W. and K.YO. (2002) The Impact of Multi-storey Car Parks on Wind Pressure Distribution and Air Change Rates of Surrounding High Rise Residential Buildings in Singapore. *International Journal on Architectural Science* 3, no. 1, pp. 30–42.

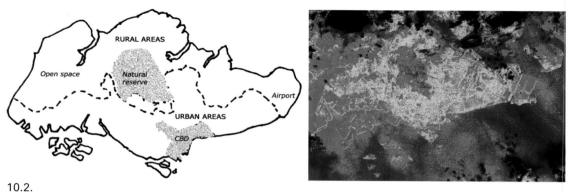

10.2.
The satellite image shows the relative temperature of the whole island during the day time derived from thermal band of Landsat-7 ETM+ on 11 Oct 2002.

exposed hard surfaces on site, such as metal roofs and runways, which will easily absorb solar radiation and incur high surface temperature during daytime. The CBD area and developments located in central southern part of the island also experience relatively high surface temperatures although they are not as high as those observed in industrial areas and the airport. Unfortunately, the magnitude of the UHI during the day cannot be generated from the satellite image directly. An intensity of 2.84°C in the daytime (1300–1500 h) was observed through island wide field measurements. Satellite images, however, are mostly obtained at noon when the intensity of the UHI may not be very obvious. To further map an islandwide temperature distribution at night when large heat island magnitude occurs, a mobile survey was conducted. Data were collected simultaneously through four survey routes which covered the whole island. The temperature mapping is shown in Figure 10.3. It can be found that lower temperatures were mostly detected at the northern part while

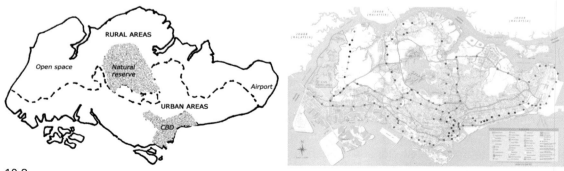

10.3.
Mapping of temperature distribution based on the mobile survey.

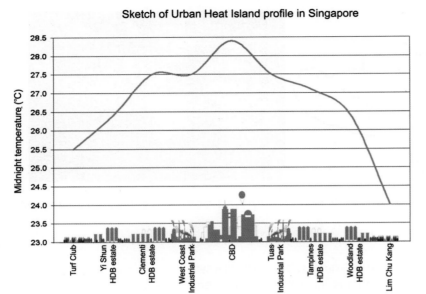

Sketch of Urban Heat Island profile in Singapore

10.4.
Sketch of Urban Heat Island profile in Singapore.

higher temperatures were observed at the southern part, especially at the CBD area. The results of the survey accorded with the partition of "urban" and "rural" areas of Singapore. Basically, the temperatures measured near to large green areas are relatively lower than those measured far away from them. A difference of 4.01°C between the urban and rural area was observed at late night (0200–0400 h).

Based on the data derived from the mobile survey, the sketch of UHI profile in Singapore was plotted (see Figure 10.4). It seems that the temperatures measured within different land uses are closely related to the density of greenery. In CBD area, absence of plants and high density of buildings caused the highest temperature. Higher temperatures were also detected in some industry areas. The possible reason may be the employment of metal roofs for industrial buildings and lack of tall trees which can cast efficient shadow on buildings. The temperatures measured within residential areas vary according to their locations. Residential areas with extensive landscape can experience lower temperatures. The open space/recreation area and the forest have the lowest air temperatures since they are very well planted.

To further explore the UHI effect over a long period, weather data for the past 20 years have been collected from the standard network of the meteorological stations in Singapore (see Figure 10.5). The yearly average, maximum and minimum temperatures of four weather stations are presented in Figures 10.6–10.9. It can be

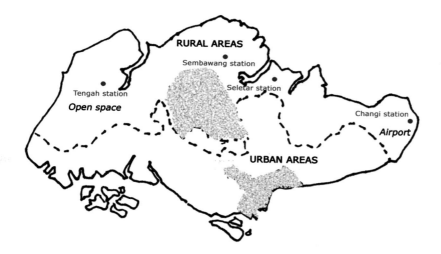

10.5.
Locations of weather stations in Singapore (Paya Lebar station is out of use).

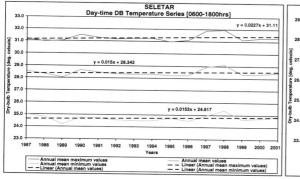
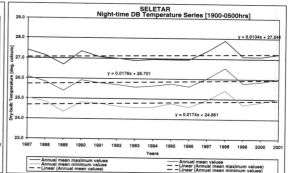

10.6.
Seletar station (day-time and night-time).

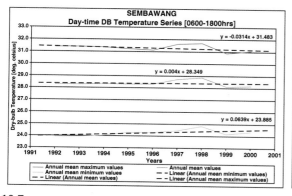
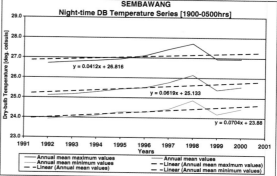

10.7.
Sembawang station (day-time and night-time).

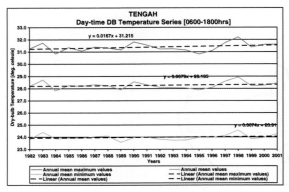

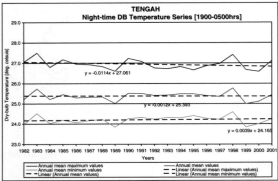

10.8.
Tengah station (day-time and night-time).

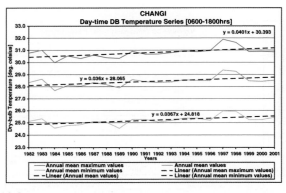

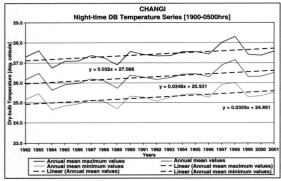

10.9.
Changi station (day-time and night-time).

observed that a clear increase in temperature over the past 20 years in weather stations at night time except in Tengah station. The possible reason for the minimum fluctuation of temperatures in Tengah station is because over 70% of land use in Tengah is open space with extensive vegetation. Less construction has been done and so influence from buildings and anthropogenic heat is minimal. On the contrary, the Changi area has the rapid increase of temperature (around 1°C) over the past 20 years. It is strongly related to the expansive construction of the airport and the increasing air traffic. The satellite image, mobile survey, and the weather data analysis all testify the occurrence of the UHI effect in Singapore. Quantitatively, a difference of 4.01°C was observed between the well planted area and CBD area.

10.3 STRATEGIES OF MITIGATING UHI

10.3.1 Urban greenery

Greening urban area is one of the main strategies in mitigating the UHI effect since vegetation plays a significant role in regulating the urban climate. Basically, plants can create an "oasis effect" and mitigate the urban warming at both macro- and micro-level. At micro-level, vegetation around buildings can alter the energy balance and the cooling energy requirements of particular buildings through sheltering windows, walls, and rooftop from the incident solar radiation and radiation reflected from surrounding buildings. As soon as a bare hard surface is covered with plants, the heat-absorbing surface transfers from the artificial layer to the living one. At macro-level, the energy balance of the whole city can be modified by adding more evaporating surfaces by throughout in the forms of natural reserves, urban parks, neighbourhood parks, and rooftop gardens. They provide sources of moisture for evapotranspiration and more absorbed radiation can be dissipated to be latent heat rather than sensible heat. As a result, the urban temperature can be reduced.

Field measurements were conducted to explore the thermal protection of rooftop gardens in Singapore. The data derived from the measurements indicate that the installation of rooftop gardens can significantly provide thermal protection to buildings and improve the surrounding environment. For example, rooftop gardens can efficiently reduce surface temperatures of roofs (Figure 10.10). Compared with a bare roof, the maximum temperature reduced by plants was observed to be around 30°C. The temperature measured under vegetation varied according to the density (LAI) of leaves. The field measurements also show lower ambient air temperatures

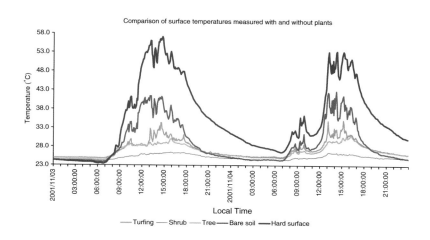

10.10.
Comparison of surface temperatures measured with different kinds of plants on a local rooftop garden on 3 and 4 November 2001.

Downloaded Data - Thursday, April 12, 2001

(TP120)-Temp°C Min 28.3 Max 41.4

(TP120)-Temp°C Min 25.9 Max 45.3

(TP120)-RHI% Min 53.9 Max 98.0

(TP120)-RHI% Min 56.6 Max 92.5

10.11.

Air ambient temperature and relative humidity plotted over 3 days on planted and bare roofs, respectively (C2 is the planted roof while C16 is the bare roof without planting).

over the vegetated areas (Figure 10.11). Throughout the day, a maximum temperature difference was observed to be around 3°C. However, plants also bring extra moisture in air, resulting in high relative humidity.

Rooftop gardens and other forms of greenery planted around buildings can greatly influence the micro-climate. However, the impact is localized compared with large urban greens. It is believed that urban parks can extend the positive impact to the surrounding built environment at a large scale. Measurements therefore were carried out in local parks and their surrounding areas. Figure 10.12 illustrates the comparison of average air temperatures measured at different locations which are lined up at a certain interval in the park and away from it in the surrounding area. From locations 1–4, which are located in the park, the average temperatures range from 25.2 to 25.5°C. There is a 400-m low-temperature region which ranges from 25.6 to 26.9°C in the surrounding residential area. The highest temperature was observed at location 9. It is 1.3°C higher than the temperature obtained at location 6 which is near to the park.

Using a location in the park as the benchmark, a computational simulation was carried out. The corresponding cooling energy savings for a typical commercial building can be up to 9%

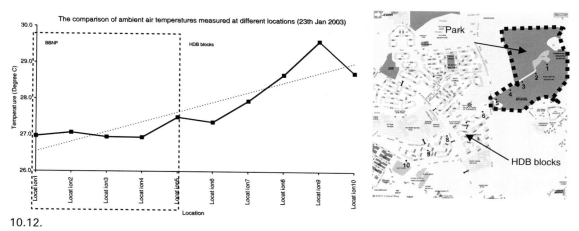

10.12.
The comparison of average air temperatures measured at different locations in the park and nearby HDB blocks (11th Jan to 5th Feb 2003).

Table 10.1. Energy savings at locations placed away from the park

	Cooling load (kWh)	Energy savings (compared with 400 m) (%)
In the park	9077	10
100 m from the park	9219	9
200 m from the park	9383	7
300 m from the park	9672	4
400 m from the park	10123	0

(see Table 10.1) when it is located near to the park within the low-temperature region.

10.3.2   Urban geometry

Urban geometry plays an important role in mitigating the UHI effect as well. One of the main reasons for the heat build up in cities is poor ventilation. However, the urban wind can be dominated and modified by urban design. The main urban design elements which can modify the wind conditions are the overall density of the urban area, size and height of the individual buildings, existence of high-rise buildings, and the orientation and width of the streets. Since CBD area is the place where serious UHI effect occurred, it has been further studied by CFD simulations (see Figure 10.13).

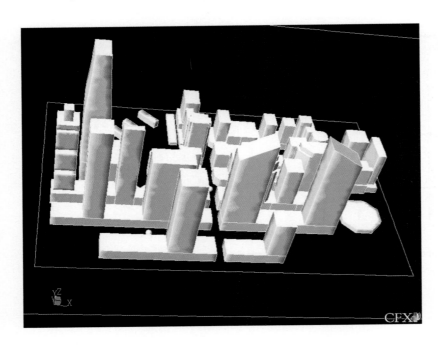

10.13.
Geometry in CFD model.

From a series of parametric studies, it can be derived that the velocities within the canyon were highly enhanced by the presence of high rise towers when the wind flow is parallel as well as perpendicular to the canyon. For parallel flow, the velocity has been increased by up to 75% and the temperature has been reduced by up to 1°C with the introduction of high rise towers. For perpendicular flow the velocity has been increased up to 800% and the temperature has been reduced by 0.8°C. The high rise towers at the canyon entrance will influence the velocities at locations deep inside the canyon length even at the lower zones. The channelling effect observed for larger H/W ratio in the absence of the high rise towers continued to stay even for smaller H/W ratio with the introduction of high rise towers. That means, unobstructed flow occurred at larger widths. This phenomenon was more prevalent at the lower zones. At the upper zone, the velocity and temperature did not change significantly in the absence of high rise towers. However, when the high rise towers were introduced, the velocity increases with increase in street width. Also the temperature decreases with increase in street width. For perpendicular flow, the velocities were much higher with the presence of high rise towers even up to one order magnitude. The transition to wake interference flow occurred at a larger H/W ratio. The velocities as well as temperatures of the lower zone were highly influenced by the high temperature of the road surface. The temperatures were much higher near the ground. But at the middle of the canyon length the introduction of high rise

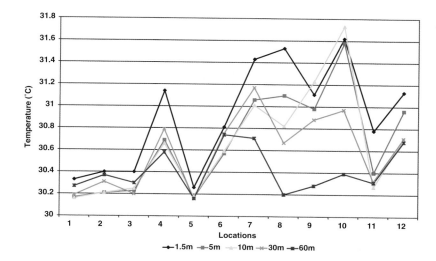

towers caused strong airflow at the lower zone thus decreasing the temperature.

The vertical distribution of temperature and velocity were rather interesting (see Figure 10.14). For most of the points, maximum temperatures were observed near the ground due to the high temperature of ground surface. Near high-rise buildings, lower and higher area of the base case had similar pattern for velocity and temperature. The temperature variation amongst the points were much lower in the higher area compared to lower area. Along the streets, high wind speeds were observed within the lower area as velocity was enhanced by the channelling effect. However, locations just in front of the high rise towers had increasing velocities with the increase in height. Though street width modification resulted in lower temperature and higher velocity at some of the locations, such a negative correlation was not obtained for all the locations. Higher wind speed did not result in lower temperatures in some cases. Near the high rise buildings, the effect of street width modification within the lower continuous canyon was different from the effect above the canyons. It was observed that an H/W ratio of 0.6–0.66 gave rise to maximum velocity at the centre of the canyon. Adapting this H/W ratio increased the velocity up to 35% and reduced the corresponding temperature by up to 0.7°C. Modifying the neighbouring street geometry enhanced the wind speed and lowered the temperature of narrow lanes. Reducing the height of the high rise towers reduced the wind speed at most of the locations and also resulted in higher temperatures. This shows that the high rise towers randomly placed above the continuous canyons did not cause any temperature increase and it is desirable to go even higher.

10.3.3 "Cold" material

Reducing the solar heat gain in an urban environment is another usable means to mitigate the UHI effect. The role of building materials, which is mainly determined by their optical and thermal characteristics, is crucial in reducing the solar heat gain. Generally, two significant factors, the albedo which is the ratio of the amount of light reflected from a material to the amount of light shone on the material and the emissivity which is the ratio of heat radiated by a substance to the heat radiated by a blackbody at the same temperature, are considered. The former factor governs the absorption of solar radiation and the later controls the release of long wave radiation.

The effect of colours has been studied through a laboratory test. The reflectance of some commonly used façade materials including aluminium panel, glass, and tile were measured. It is obvious that the reflectance changes with the variation of the colour of the façade material. For the same façade material (while other parameters remain the same), the lighter the colour of the material, the higher the reflectance. In another words, the material with lighter colour can effectively reflect incident solar radiation. The laboratory test results of 9 different colours of aluminium are listed in the Table 10.2. It can be observed that reflectance varies greatly from 83.03 to 4.36% when the painted colour changes from white to black.

Table 10.2. The property of aluminium with different colour

Colour of aluminum panel	Reflectance in the UV wavelength range (%)	Reflectance in VIS wavelength range (%)	Reflectance in NIR wavelength range (%)	Solar reflectance (%)
White	9.51	86.17	87.59	83.03
Lemon	3.13	49.04	86.65	65.17
Silver	52.49	54.74	67.68	60.97
Grayblue	33.53	41.51	61.28	50.8
Red	3.1	17.37	80.83	47.76
Golden	4.69	33.76	63.79	47.02
Gray	16.31	19.51	38.37	28.59
Darkblue	4.48	8.00	47.26	27.06
Black	1.29	2.47	6.46	4.36

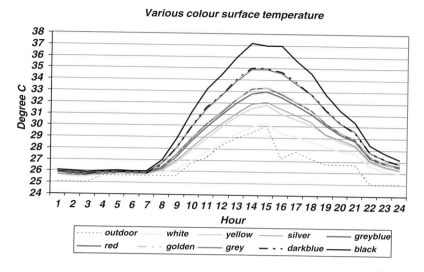

Various colour surface temperature

Degree C

Hour

Legend: outdoor · · · · · · · white ——— yellow ——— silver ——— greyblue ——— red ——— golden – – – grey ——— darkblue — · · — black ———

10.15.
Various surface temperatures caused by different colour of building material.

The reflectance governs the external surface temperature of façade. Through a simulation program (TAS), the effect of reflectance on surface temperature of façade under the local weather condition was worked out. Figure 10.15 shows the simulation results of the surface temperatures of the various façades with different colour (reflectance) in a typical day in Singapore. It indicates that the dark colour buildings have higher external surface temperatures compared with those light colour ones. In the local tropical climate, the maximum difference of the external surface temperature can be easily 7°C.

Modelling the CDB area using CFD simulations have shown that façade materials and especially their colors play a very important role in the determination of the thermal environment inside the urban canyons. At higher wind speeds, the effect of materials on the air temperature was not significant for wider canyons. However when low albedo materials were used, narrow canyons situated away from the inlet had higher temperature even at higher inlet wind speeds. The temperature in the middle of such canyon was increased by 1°C. At very low inlet wind speeds, the effect of materials was found to be very significant and the temperature at the middle of the narrow canyon increased up to 2.52°C with the façade material having lowest albedo. It was noted that the temperature at the middle of the wider canyons also increased up to 1.3°C. The increase in temperature inside the canyons can cause the peak electricity load to increase by up to 6.8–9%.

The so-called "cold" materials, which are characterized by high reflectivity and high emissivity, can improve thermal conditions in cities through contributing lower surface temperatures. The benefits of employing the "cold" materials in cities, especially in tropical cities with very strong solar radiation, are obvious. First, the "cold" materials can mitigate the UHI effect and reduce the cooling energy use in cities. Second, the "cold" materials can affect the generation of smog. Finally, the "cold" materials also have longer lifetimes because they are not as stressed by the excessive heat.

10.4 CONCLUSION AND RECOMMENDATION

The research uncovered the severity of UHI effect in Singapore through satellite images, mobile survey, and weather data analysis. It indicates a UHI intensity of about 4°C can be observed. The UHI effect does occur in the country. Urban greenery, "cold" materials, and urban geometry are three main measures considered to mitigate the existing UHI effect in Singapore.

Vegetation can improve the urban thermal environment from both macro and micro perspectives. At the macro-level, large city greens benefit their surroundings. A low-temperature region can be formed near to a large park. At the micro-level, vegetation strategically placed around roofs and walls can be considered as a complement of urban greens. It is found that, with the shading of plants, surface temperatures measured under different kinds of vegetation were much lower than that measured on the hard surface. The temperature measured under vegetation varied according to the density (LAI) of plants. Normally, lower temperatures were measured under thick foliage while higher temperatures were obtained under sparse foliage or only soil. The heat transfer through the bare roof was greater than that through planted roofs and roof with only soil. Less solar heat reflected by the greenery and less long wave radiation emitted from the planted roof were confirmed through comparisons of reflected solar radiation, global temperatures, and Mean Radiant Temperatures (MRTs) measured on site.

The parametric studies furnished some insights into the effect of high rise towers on the airflow and temperature profiles within urban canyons. The velocities were highly enhanced by the presence of high rise towers. Also the temperature was reduced by up to 1°C. The series of CFD simulations modelling the CBD area provided an understanding of the complex airflow and heat transfer phenomena occurring in the urban centre. It can be concluded that the velocity was highly enhanced by the channelling effect as well as presence of high rise towers. The thermal environment was found to be influenced by the H/W ratio of the continuous canyons. Adapting an optimum H/W ratio increased the velocity up to 35% and reduced the corresponding temperature by up to 0.7°C at the middle of the road. The existing high rise towers placed randomly above the continuous canyons in fact facilitated to increase the airflow and lower the air temperature.

The selection of material is significant for the thermal performance of building façades and the urban thermal environment. A material with high albedo can reduce the solar heat gain during daytime. The surface temperature of the material is lower than that of a material with low albedo. Since the urban ambient temperature is associated with the surface temperatures of the building façade, lower surface temperature can obviously help decreasing the ambient air temperature and eventually contribute to better urban thermal environment.

The UHI effect is not solely found in Singapore but other tropical cities like Kuala Lumpur, New Delhi, and so on (Santamoouris, 2002, pp. 56–57). The three basic strategies, therefore, are applicable to all cities facing the negative impacts of UHI. But the three strategies may not be equally important and the judgment is very much related to the local condition of every individual city. Based on the findings delivered from our study (Wong, 2003), the following specific recommendations are made for mitigating the UHI in Singapore:

- Through the satellite image, the "hot" spots are normally observed on exposed hard surfaces in the urban context during the daytime. It is suggested that these exposed hard surfaces should be strategically shaded by greenery or artificial sun-shading devices.
- Historical analysis of long-term climatic data of Singapore indicates the raise of temperature is associated with the land uses. It is believed that implementing greening of Singapore and minimizing the release of anthropogenic heat can mitigate the UHI effect at macro-level.
- Temperature mapping surveys show the temperatures of the developed areas are associated with the greenery coverage within the sites. The well planted areas have lower temperatures while locations with less greenery incur higher temperatures.
- The further exploration on the greenery further indicates the positive impacts of plants on mitigating the UHI effect in Singapore. It is strongly recommended that plants cannot only be introduced into a developed site as a cooling buffer but also be introduced into buildings as an insulating layer. The greenery can be introduced into the built environment in the forms of parks, rooftop gardens, and vertical landscaping.

- Through the laboratory testing and the simulations, it indicates that the colours of building materials have significant impacts on surface temperatures which subsequently influence ambient temperatures. It is suggested that more light-color materials should be employed to save cooling energy and mitigate the UHI effect.
- It was found that the heat from the asphalt road surface contributes much on the temperature increase inside the canyons. The high rise towers randomly placed above the continuous canyons in fact enhance the airflow and help to reduce the temperature inside the canyons.
- Façade materials and especially their colours play a very important role in the formulation of the thermal environment inside urban canyons. At very low wind speeds, the effect of materials was found to be significant and the temperature at the middle of the narrow canyon increases significantly with the façade material having low albedo.

REFERENCES

Bay, J.H.P. (2001) Three Tropical design paradigms, in Tzonis, A., Lefaivre, L. and Stagno, B. (eds), *Tropical Architecture–Critical Regionalism in the Age of Globalization*. Great Britain, Wiley-Academy, pp. 229–265.

Bridgman, H., Warner, R. and Dodson, J. (1995) *Urban Biophysical Environments*. Melbourne, New York: Oxford University Press.

Landsberg, H.E. (1981) *The Urban Climate*. New York: Academic Press.

Nichol, J.E. (1994) Modelling the relationship between LANDSAT TM Thermal Data and Urban Mornholoev. *Proc. ASPRSIACMS Annual Convention and E;positi*.

Nieuwolt, S. (1966) The urban microclimate of Singapore. *The Journal of Tropical Geography*, 22, pp. 30–31.

Oke, T.R. (1978) *Boundary Layer Climates*, William Clowes and Sons Limited, London.

Oke, T.R. (1987) *Boundary Layer Climates*. (2nd edn.), New York: Methuen.

Roth, M. and Oke, T.R. (1989) Satellite-drived urban heat islands from three coastal cities and the utilization of such data in urban climatology, *International Journal of Remote Sensing*, 10, no 11, pp. 1699–1720.

Sien, C.L. (1970) Temperature and humidity observations on two overcast days in Singapore. *Journal, Singapore National Academy of Science*, 1, no 3, pp. 85–90.

Santamoouris, M. (ed.) (2002) *Energy and Climate in the Urban Built Environment*. London: James and James Science Publishers.

Tso, C.P. (1996) A Survey of Urban Heat Island Studies in Two Tropical Cities. *Atmospheric Environment,* 30, pp. 507–519.

Wong, N.H. (2003) *Study of Urban Heat Island in Singapore (Research Report)*. BCA & NUS, Singapore.

11 TROPICAL URBAN STREET CANYONS

Elias Salleh

Department of Architecture, Universiti Putra Malaysia

Abstract

Urban canyons represent important urban open spaces in between buildings, having various microclimates which influence the resulting thermal environment, including the formation of "cool islands". Thermal environment is of prime importance, influencing people's use of urban outdoor spaces. Generally, a deeper urban canyon will result in less penetration of direct solar radiation to the street level, while reducing the sky radiant field and increasing air flow within the urban canyon. These are potentially favourable for mitigating outdoor thermal discomfort in highly developed tropical urban areas, and for promoting better utilisation of urban spaces for outdoor activities. An attempt to study Kuala Lumpur urban street canyons has been made, using Fanger's PMV comfort index and Terjung's URBAN3 climate model. The study confirms that a shallow urban canyon is warmer than a deeper one, and that shallower urban canyons experience higher cumulative energy fluxes than deeper ones. In shallower urban canyons small increases of air velocity have little influence on thermal comfort level. On the other hand, a deeper urban canyon with lower air velocities can maintain tolerable PMV levels, mainly because of the cooling effect from the reduction in solar penetration to the street level and the reduced sky view. For optimum shading the best street orientation for urban canyons in such locations is north/south, and the northeast/southwest and northwest/southeast orientations are good compromises. An urban canyon height/width ratio of 3:1 represents the threshold of optimum urban canyon shading and surface temperature control. These design guides are critical for designing better urban environments.

Keywords

Urban open spaces, urban canyons, street activities, street shading, climate modelling, thermal comfort.

11.1   INTRODUCTION

Cities are generally made up of densely developed built forms, mainly buildings, with variable active and passive open spaces in between them. These urban open spaces comprise large open spaces, for instance urban parks, and smaller spaces in between buildings to which the term "urban canyon" is normally applied. This article discusses the urban street canyon as an important element of urban design in the tropics, and focuses on its micro-climatic impact on the thermal environment. The latter involves quantitative assessment of the parameters affecting the thermal environment of tropical urban street canyons with specific reference to Kuala Lumpur. The comfort index used was the Predicted Mean Vote (PMV) developed by Fanger (1970) while an existing urban climate model URBAN3, originally developed by Terjung and Louie (1974), was used to determine the surface energy budgets, surface temperatures of urban canyons and building systems, and urban canyon shading. It is hoped that the findings from this study can beneficially inform the process of designing appropriate urban street canyons in the tropics.

11.2   URBAN CANYON

The urban outdoors is generally perceived as an open and unob-structed space for vehicular traffic and pedestrian movement. It consists of spaces that are geometrically bounded by a variety of elevations. In spatial terms Krier (1984) has classified these urban spaces into two distinct forms, namely the "square" and the "street" – analogous to the "room" and the "corridor" of interior spaces.

The square as classified by Krier can be perceived as enclosed or partially enclosed space, represented by such spaces as inner court-yards or atriums in the private sphere, and market squares, city-hall squares, mosque squares, ceremonial squares, agora and parks in the more public sphere. Functionally, the square is regarded as a communal focal point for gathering, meeting, outdoor recreation, resting or pausing. The term "plaza" is also commonly used to refer to such spaces, meaning a place.

The street, on the other hand, is a more functional space pro-viding the framework for access and movement to all parts of a city. It is part of a dynamic continuum of spaces represented by alleyways, corridors, arcades, malls, high streets, boulevards and avenues. Despite the increasing numbers of and popularity of indoor air-conditioned malls, the street remains an important part of urban life, providing opportunities for many outdoor commercial and

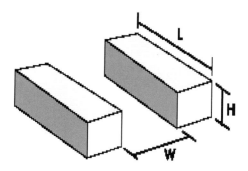

socio-cultural activities. In fact, due to the close inter-relationship of buildings and streets through perceived continuity of volume and inter-dependent mixes of elements and functions, the street tends to act both as an extension of interior spaces and as exterior rooms in the city. In this article the term "street" is used specifically to refer to the "urban street canyon" (Figure 11.1).

11.3   STREET ACTIVITIES

Street activities may range between two extremes: one being dominated by free-flowing vehicular traffic, and the other where vehicular traffic is subordinated to pedestrian needs and environmental factors. Activities related to pedestrian needs can be categorised into three groups: circulation uses, commercial uses and social-plus-amenity uses.

For circulation uses the street is treated as a transient space, an interchange, or a linkage for waiting, boarding and alighting from vehicles, for crossing streets, or for entering and leaving subways, buildings or overhead bridges. In terms of commercial uses streets have been traditionally, and still are, the natural venues for street vending, hawking, peddling, bargaining and window-shopping. Bazaar or open-market activities, which enliven the street continuously or at regulated times, are common in South-East Asian cities. Sidewalk cafés which provide for outdoor eating, drinking, relaxing, people-watching and waiting, are a metropolitan feature that has been especially important in European streets but has also gained popularity in South-East Asian streets. Additionally, retail shops, arcades, restaurants and banks are as much a part of the street on which they front as the street itself. All these activities would naturally be located near the edges of streets, where protected corridors or "five-foot ways" of buildings meet the externally exposed street.

As public spaces, streets can be used for strolling, hanging-out or queuing, while works of art, banners and street performers may grace some main streets. Finally, annual parades, celebrations, and carnivals add much fun and excitement to the major streets of cities.

The particular experience of a single street, or a collection of streets and the activities, buildings and other sights along them will contribute to the image of a city. This is emphasised by the fact that streets are by far the most common and busy urban open spaces, occupying substantial proportions of a city area.

11.4   STREETS OF KUALA LUMPUR

In Kuala Lumpur, the capital of Malaysia, the most significant street activities relating to traditional commercial practices can be found in Jalan Petaling (Figure 11.2), Jalan Tuanku Abdul Rahman, Jalan Masjid India (Figure 11.3) and Jalan Chow Kit. These four locations have been synonymous with the city itself since its early days, each having a distinctive identity relating to the multi-ethnic nature of Malaysian society. Recent developments have, however, changed the whole physical look of some of these streets. "Street roofs" have been built over them, in the belief that these structures would provide relief the climate. While the "street roof" in Jalan Petaling seems to have positive effects on street activities, the one in Jalan

11.2.
Jalan Petaling: the Chinese image and "street roof".

11.3.
Jalan Masjid India: the controversial "street roof".

11.4.
Jalan Bukit Bintang: a cosmopolitan image.

Masjid India, at the time of writing, has not achieved the same effect and acceptance, either functionally or architecturally.

In keeping with global urban development trends, certain streets in Kuala Lumpur have evolved commercially to have more cosmopolitan characteristics. The most prominent among these is Jalan Bukit Bintang (Figure 11.4), which has managed to shed its formerly less savoury reputation for a more-respectable cosmopolitan image, thus making it more attractive to foreign and local patrons alike. The other area that has come into prominence is Jalan P. Ramlee which stretches from the Kuala Lumpur City Centre (which houses the Petronas Twin Towers) to the busy Jalan Sultan Ismail (which is an established "hotel corridor"). These one-way streets offer restaurants, sidewalk cafés and covered outdoor sitting areas, while leading into the entrances of prominent shopping malls, hotels and office complexes.

It should be noted that, while the more "conservative" streets of Jalan Petaling and Jalan Masjid India have been roofed-over with large "street roof structures", Jalan Bukit Bintang and Jalan P. Ramlee have smaller "sidewalk roof structures".

11.5   STREET MICROCLIMATE

The urban climate has been the subject of great interest, due to the rapid urbanisation process that is taking place globally and the ever-growing concern to have a sustainable environment. Since its discovery by Luke Howard in the early 19th century, the "urban heat island" (UHI) has been extensively studied and many researchers have written on this subject including Oke (1988) and Sham (1987). There have also been many studies of the urban climate from energetics and thermal comfort perspectives. These include those of Burt et al. (1982); Todhunter and Terjung (1988) and Arnfield (1990). These and many other related studies have the common objectives of investigating the effects of urban morphology on urban climate and the consequent thermal stress on man.

Tropical climates are generally dominated by high daytime air temperatures throughout almost the whole year, giving rise to outdoor thermal discomfort. The ability to moderate outdoor thermal stress by eliminating or reducing extreme conditions would certainly benefit the users of urban outdoor spaces. This would also indirectly benefit indoor conditions, as it would mean reducing the stresses on buildings by moderating adjacent microclimates.

Urban canyons, represented by plazas and streets, have varied microclimates giving rise to a variety of thermal environments, including "cool islands" (areas that are cooler than the typical surroundings) within the urban canopy layer. The occurrence of cool islands in tropical cities generally results from a combination of many factors, including the existence of green spaces and vegetation as well as the shading of outdoor spaces by buildings or other built forms. Shading of streets due to urban canyon geometry is an inherent property of street canyons that needs to be properly understood to mitigate the thermal environmental stress.

11.6 BASIC STREET SHADING

Street activities in most climates are partly affected by sunlight availability or the lack of it. While direct sunlight is very welcome in cold climates, the same does not apply in hot weather conditions, especially in tropical climates. Areas of natural shade arising from site topography, trees or even built forms are thus regarded as positive elements in the tropical climates, which should be taken advantage of. In this respect, shading of street edges or sidewalks is an important urban design goal.

In the urban context the street is bounded by two vertical building façades fronting the street, with each façade obstructing direct sunlight on the opposite street edge. This is referred to as "obstruction angle" or "the spacing angle" by Evans (1980), and, in a street canyon system, is represented by the vertical angle subtended by the top of a building to the base of the opposite building, or conversely the angle from the horizon to the top of the obstructing building observed from the base of the opposite building or the street edge. At this angle (α) the street surface is fully shaded from direct sunlight. This angle can be obtained as follows:

$$\tan \alpha = H/W$$

where α = the obstruction angle, H = the height of the opposite building and W = the distance between the buildings or the street width.

Table 11.1 illustrates the equivalent "obstruction angle" for some H/W ratios. Hence a street with an H/W ratio of 1:1 has an α value of 45°, and a street with an H/W ratio of 3:1 has an α value of 71.6°.

Expressing the street geometry in terms of "obstruction angle" (α) is perhaps more appropriate to the designer than in terms of H/W

Table 11.1. H/W ratio and obstruction angle equivalence

H/W ratio	Obstruction angle
0.50:1	26.6°
1.00:1	45.0°
1.50:1	56.3°
2.00:1	63.4°
2.50:1	68.2°
3.00:1	71.6°
3.50:1	74.0°
4.00:1	76.0°

Table 11.2. Solar vertical shadow angles for latitude 4°N

Canyon orientation	Time													
	10.00 a.m.							3.00 p.m.						
	Dec	Jan Nov	Feb Oct	Mar Sept	Apr Aug	May Jul	Jun	Dec	Jan Nov	Feb Oct	Mar Sept	Apr Aug	May Jul	Jun
N/S	59°	59°	59°	60°	60°	60°	61°	43°	43°	44°	44°	45°	45°	45°
NW/SE	89°	86°	76°	68°	63°	57°	54°	38°	40°	46°	52°	60°	68°	72°
NE/SW	49°	52°	58°	65°	73°	80°	84°	90°	85°	76°	69°	63°	57°	55°
E/W – north facing	–	–	–	–	84°	70°	67°	–	–	–	–	80°	87°	62°
E/W – south facing	59°	63°	75°	86°	–	–	–	54°	58°	72°	86°	–	–	–

ratios, since the angle values can be directly compared to those of relevant solar angles at various times of the day and year for assessing street canyon self-shading capacity.

For this purpose, the solar vertical shadow angle (SVSA) is used to represent the projection of the solar altitude angle (SAA) onto an imaginary plane perpendicular to a street canyon face. When the sun is directly opposite the street canyon face, SAA and SVSA are identical; for all other solar positions SVSA values differ from SAA values. Table 11.2 illustrates the various solar vertical shadow angles (SVSA) at 10.00 a.m. and 3.00 p.m. for four street orientations at latitude 4°N (derived from the relevant solar chart).

As an example, a street with an H/W ratio of 3:1 ($\alpha = 71.6°$) oriented NE/SW will not be fully protected from the 3.00 p.m. sun from October through to February (SVSA = 76°–90°), but will be fully protected for the other months (SVSA = 55°–69°).

Establishing street canyon self-shading characteristics, as described above, is important in understanding the potential for solar penetration into a street canyon. Additional measures then can be designed to further moderate the thermal conditions in the street with a full understanding of the architectural context, technology and microclimatic consequences.

Currently, many architectural design software packages have the facility interactively to investigate shadow casting by buildings, in graphic and digital form; among the more user-friendly ones are *Ecotect* (Figure 11.5) and *SketchUp*. These design tools can address the issue quite simply and while a design is being developed.

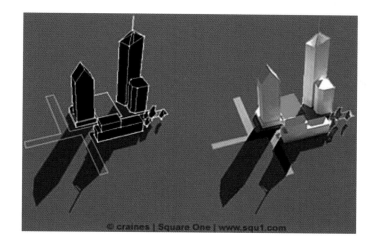

11.7   FIELD MEASUREMENT PROGRAMME

A field measurement programme and computer simulation study of the urban microclimate have been conducted in Kuala Lumpur as part of a higher degree requirement (Salleh, 1994). The following section of this chapter will describe in detail some of the findings of the study, which can assist in better understanding the thermal environment of a street in a tropical city.

11.7.1   Aim and methods

The aim of the study was to seek proof that in a warm–humid climate, the urban microclimate is sensitive to urban morphology, and that despite the unfavourable perceptions normally associated with high-rise development, the resulting outdoor thermal environment in an urban canyon can create tolerable conditions for humans. The study involved the following two key tasks:

- Conducting a climatic measurement study of two separate urban canyons in Kuala Lumpur;
- Evaluation of urban canyon shading, surface temperatures and the total radiant energy fluxes.

11.7.2   Damansara and Melawati Street Canyons

Two sites representing two typical Asian urban street canyons were selected, namely Damansara Town Centre and Melawati Commercial Centre.

Damansara is a commercial development comprising five- to seven-storey office buildings, typifying the physical characteristics of street canyons found in the more developed city centres of many tropical Asian cities, such as Hong Kong, Singapore, Bangkok, Jakarta and Kuala Lumpur. It is a development where the street activities would be related to the retail shops, arcades, banks and restaurants fronting the streets.

Melawati, comprising mainly two-storey shophouses, on the other hand would typify the common street canyons of the majority of urban developments in many tropical Asian countries, where commercial and social-plus-amenity uses would prevail. These involve mainly low-to-medium rise developments.

The street canyons in both sites are aligned north/south. The former has a canyon height-to-width (H/W) ratio of 4:1 (west side) and 3:1 (east side), while the latter has a symmetrical H/W ratio of 1:1.

11.7.3   Measurement programme

The objectives were to sample primary data on urban microclimate, PMV values and urban surface temperatures of two selected sites in Kuala Lumpur (Figures 11.6 and 11.7).

The measurements were carried out in June and July 1987 in the then newly completed (but unoccupied) Damansara Town Centre (Damansara) and Melawati Commercial Centre (Melawati).

11.6.
Damansara site (now).

11.7.
Melawati site (now).

Measurements were taken using a thermal comfort meter (B&K type 1212), indoor climate analyser (B&K type 1213), silicon pyranometer (Haenni Solar 118&130), solarimeter, Rion digital anemometer, globe thermometer and thermocouple wires connected to a data logger and IBM PC as well as a Stevenson Screen. The Damansara site was subjected to 24-h measurement on June 28–29 and from 8 a.m. to 7 p.m. on June 26 and 27. The Melawati site was subjected to 24-h measurement on July 2–3 and from 8 a.m. to 7 p.m. on July 28 and 29. Reference weather data from the nearby Petaling Jaya Meteorological Station were used. While global solar radiation and surface temperatures were continuously recorded by the data logger, other readings were recorded manually on an hourly basis.

The observations and conclusions derived have not included normalisation to the measured data because of the different measurement dates. The PMV data, however, do reflect the resulting integrated values.

11.7.4 Air temperature

Higher daytime air temperatures existed for longer in Melawati (over 30°C from 10.30 a.m. to 7.30 p.m.) compared to those in Damansara (over 30°C from 12.00 p.m. to 5.00 p.m.). This seems to reflect the fact that a more open and exposed urban canyon will result in longer hours of high daytime air temperature within it. This pattern was also reflected in the Mean Radiant Temperature (MRT) readings.

11.7.5   Relative humidity

Relative humidity (RH) values below 70% were recorded for Damansara between 11.00 a.m. and 4.00 p.m., while for Melawati they were recorded between 10.00 a.m. and 6.00 p.m. The lowest RH reading for Damansara was 53% between 1.00 p.m. and 2.00 p.m., and the lowest reading for Melawati was 50% at 2.00 p.m. Thus Melawati canyon experienced longer periods of lower RH.

11.7.6   Air velocity

Air velocities were generally low within Damansara, averaging about 0.4 m/s between 8.00 a.m. and 2.00 p.m. and 0.25 m/s between 3.00 p.m. and 10.00 p.m. Air velocities within Melawati were slightly higher, averaging about 0.3 m/s between 8.00 a.m. and 12.00 p.m., 0.7 m/s between 1.00 p.m. and 5.00 p.m., and 0.4 m/s between 6.00 p.m. and 10.00 p.m.

   Thus Melawati experienced relatively higher air velocities than Damansara. However, this did not have significant effects on the PMV values.

11.7.7   Surface temperature

Measurement of ground surface temperatures was given emphasis because of the ground's close proximity to pedestrians. Readings from both sites indicated that road tarmac surfaces registered the highest temperatures at 2.00 p.m. at both sites, reaching over 45°C. Table 11.3 lists the maximum temperatures recorded.

11.7.8   PMV

PMV values were recorded almost simultaneously with the measurement of individual climatic parameters with fixed metabolic rate (MET) values of 1.2 and clothing (CLO) values of 0.6. Only PMV values of 2.0 and below were of relevance to this study.

   Calculated PMV values were derived using a computer program provided in ISO 7730-1984. The inputs for the calculations were taken directly from the field readings, except for the Mean Radiant Temperatures (MRT) which were derived from Plane Radiant

Table 11.3. Maximum surface temperatures (Salleh, 1994)

Location	T_{max} (°C)	Time of day
Damansara		
Tarmac (west)	48.5	2.00 p.m.
Tiled floor (canyon centre)	37.8	1.00 p.m.
Tiled floor (west outer corridor)	41.2	3.00 p.m.
Melawati		
Tarmac (west)	47.6	2.00 p.m.
Concrete slab (canyon centre)	44.2	2.00 p.m.
Tiled floor (canyon, west-facing corridor)	40.6	3.30 p.m.

Temperature (Tpr) readings using a radiant temperature asymmetry transducer (one of five transducers of the indoor climate analyzer).

Over the period of measurement it was observed that PMV values for the inner corridors in Damansara managed to remain below 2.0 (warm) throughout the day, except for the canyon centre which registered PMV values higher than 2.0 between 12.30 p.m. and 3.00 p.m. The inner corridors in Melawati exceeded a PMV value of 2.0 from around 12.00 p.m. to around 5.00 p.m. The canyon centre registered PMV values of over 2.0 between 10.30 a.m. and 5.00 p.m.

The daytime PMV values were generally lower in Damansara than in Melawati. However, the nighttime PMV values in Damansara remained above 1.0 until about 11.00 p.m., while those in Melawati quickly dipped below 1.0 before 8.00 p.m.

11.7.9 Comparison of measured and calculated PMV

measured PMV readings appeared to be marginally lower than calculated PMV values. Similar findings have been reported by Spain (1986) who concluded that measured PMV values were 0.33 lower than PMV values derived from individually measured parameters. Table 11.4 summarises the major measured conditions for both sites on typical days.

Table 11.4. Summary of measured conditions

Conditions	Damansara (Time of day)	Melawati (Time of day)
Air temperatures above 30°C	12.00 a.m.–5.00 p.m.	10.30 a.m.–7.30 p.m.
Relative humidity below 70%	11.00 a.m.–4.00 p.m.	10.30 a.m.–6.00 p.m.
Mean radiant temperatures above 30°C	12.30 a.m.–3.30 p.m.	10.30 a.m.–5.00 p.m.
Predicted mean vote above 2	12.30 a.m.–2.30 p.m.	10.30 a.m.–5.00 p.m.

11.8 COMPUTER SIMULATION USING URBAN3

An existing urban climate model, URBAN3, has been applied to estimate the energy budgets in the Damansara and Melawati urban canyons. URBAN3 is an Urban Canopy-Layer (UCL) model developed by Professor W.H. Terjung and his co-researchers in the 1970s (Terjung and O'Rourke, 1980), specifically to simulate the urban climate on a scale at which people occupy the urban landscape.

The model facilitates the analysis of surface energy budgets and surface temperatures of urban canyons and building systems on a block-by-block basis. Despite its steady-state nature, the model can be used with reasonable confidence to determine the energy budgets and surface temperatures of exterior building surfaces (Todhunter and Terjung, 1988).

URBAN3 consists of three major parts:

- Obstructions and view factors,
- Radiation interception,
- Temperature and energy budget calculations.

URBAN3 (as described in Terjung and O'Rourke, 1980) approximates the physical structure of a city by using a finite number of rectangular blocks intermixed with streets, parking lots and parks. The properties, dimensions and distributions of the blocks are specified to approximate the structural and material characteristics of the city or the part of the city under consideration. The blocks are translated into three-dimensional coordinates, which are used in a separate program called OBSTRUCT to generate the distances and heights of all possible obstructions in the neighbourhood.

11.8.1 Urban canyon shading

The shading of an urban canyon can be measured as a percentage of the total canyon surface area. The greater the shading percentage the less canyon surface area is exposed to sun radiation. Table 11.5 lists the shading percentages for an urban canyon (average values from 9.00 a.m. to 3.00 p.m.) with varying canyon H/W ratios at different orientations for the month of September.

For urban canyon shading in Kuala Lumpur, N/S and NW/SE street alignments have an edge over other orientations. These orientations can provide more than 50% shading if the canyon H/W ratio is 3:1 and above. On average they can provide between 5 and 15% more shading than other orientations, regardless of canyon H/W ratio. Thus for N/S orientation Melawati has 21% less shading than Damansara.

Table 11.5. Shading percentages for various urban canyon orientations and H/W ratios (Salleh, 1994)

H/W ratio	Shading percentage for various orientations (%)			
	E/W	NW/SE	N/S	NE/SW
1:1*	27.9	30.7	32.0	26.6
2:1	36.5	44.0	45.1	38.7
3:1**	42.6	52.3	53.4	47.5
4:1**	46.8	58.0	59.3	53.0
5:1	49.6	63.4	64.4	57.4

*Applicable to Melawati
**Applicable to Damansara

11.8.2   Street surface temperature

URBAN3 calculates surface temperatures of sunlit and shaded portions of the street separately. The temperatures were averaged to produce a single representative value. In the assessment of street surface temperatures the midday period (i.e. 11.00 a.m. to 1.00 p.m.) was omitted because self-shading during this period cannot be achieved by building blocks. Hence, only the average surface temperatures between 8.00–10.00 a.m. and 2.00–4.00 p.m. were considered, as shown in Table 11.6.

The results indicate that an urban canyon aligned along the E/W orientation will have higher average street surface temperatures than any other orientation. Given similar conditions, street surface temperatures for an *H/W* ratio of 1:1 exceed those for an *H/W* of 3:1 by at least 10°C on the average.

Table 11.6. Calculated urban canyon surface temperatures (Salleh, 1994)

H/W ratio	Average surface temperature for various orientations (°C)			
	E/W	NW/SE	N/S	NE/SW
1:1*	38.9	31.3	31.4	37.5
2:1	33.2	21.8	22.1	25.4
3:1**	29.6	20.5	20.9	21.2
4:1**	28.8	20.1	20.3	20.5
5:1	28.1	20.1	19.8	19.6

*Applicable to Melawati
**Applicable to Damansara

Table 11.7. Total potential urban canyon radiant heat
(Salleh, 1994)

Type	Damansara	Melawati
(i) Total system short-wave radiation	2.7	4.5
(ii) Total terrestrial long-wave radiation	8.4	5.3
(iii) Total sky radiation	5.0	7.5
(iv) Net long-wave radiation	2.0	3.2
(v) Total system re-radiation	15.5	15.9

11.8.3 Total potential radiant heat

URBAN3 calculations of energy budgets in Damansara and
Melawati, based on clear sky conditions, resulted in the cumulative
quantities (simplified to cal cm^{-2} d^{-1}) shown in Table 11.7.

For incoming long-wave radiation the main determinant is canyon
geometry and the rate of radiation received is fairly constant for all
orientations. On the other hand long-wave re-radiation output does
vary with orientation, similar to the surface temperature patterns in
Table 11.6. The results point to an optimum canyon H/W ratio of 3:1.

The results also indicate that Melawati canyon receives 66%
more short-wave radiation, 50% more sky radiation, and 60% more
net long-wave radiation than Damansara canyon, while the later
receives 58% more terrestrial long-wave radiation. Total system
re-radiations are, however, similar.

Summing items (i) to (iv) will result in 18.1 cal cm^{-2} d^{-1} for
Damansara and 20.5 cal cm^{-2} d^{-1} for Melawati. Therefore, cumu-
latively, Melawati canyon receives 13% more energy fluxes than
Damansara canyon.

11.9 SUMMARY OF FINDINGS

The study of street microclimates in Kuala Lumpur has confirmed that urban microclimatic conditions in a warm–humid climate
are sensitive to urban canyon geometry and street orientation.
Field measurements have established that a shallow urban street
canyon is warmer than a deeper one in the daytime, but during
the night it can cool more rapidly than the deeper canyon, which
remains warmer for much longer.

Energy budget simulations of representative sites in Kuala
Lumpur confirm that the shallower urban street canyon experiences higher cumulative energy fluxes than the deeper urban street
canyon – in this case 13% more. The study has also established that

small increases of air velocity associated with shallower urban street canyons have little influence on thermal comfort level. On the other hand, a deeper urban street canyon even with lower air velocities can maintain tolerable PMV levels due to reduction in solar penetration and reduced sky view. Hence, contrary to popular belief, these findings imply that there are some benefits to be derived, in terms of pedestrian thermal comfort, from having deeper urban canyons in warm–humid tropical climates, i.e. by having buildings close to each other.

The study confirms that the best orientation for urban street canyons in Kuala Lumpur is N/S, and that diagonal orientations (NW/SE or NE/SW) are a good compromise. It has also been calculated that an urban canyon *H/W* ratio of 3:1 represents the threshold for optimum urban street canyon shading and control of street surface temperatures, beyond which further height increases will yield minimal improvement.

11.10   CONCLUSIONS

The street is an important urban design element that plays a vital role in city life. Activities within a tropical street can be comfortably carried out if thermal stress can be reduced by creating "cool island" conditions. The main objective is the mitigation of thermal effects, first by capitalizing on the potential of street geometry to provide sun-shading, and second by reducing or preventing the heating-up of street surfaces. The unpredictability and normally low velocity of air movement within cities makes it less of an option.

The findings in this study have enriched the understanding of the street microclimatic implications of urban built form. This study has revealed the potential of tropical urban street canyons in high-rise developments to serve as functional and sustainable urban spaces for both commercial and social purposes. A street canyon *H/W* ratio of 3:1 has been identified as a good guideline for effective street shading and surface temperature control in tropical conditions. Additional measures, such as appropriate planting, street roof structures and sidewalk roof structures would greatly enhance the quality of street life in tropical cities.

REFERENCES

Arnfield, A.J. (1990) Street Design and Urban Canyon Solar Access. *Energy and Building* 14, pp. 117–31.

Burt, J.E., O'Rourke, P.A. and Terjung, W.H. (1982) View Factors Leading to the Simulation of Human Heat Stress and Radiant

Heat Exchange: an Algorithm. *Arch. Met. Geoph. Biokl.,* Ser. B, 30, pp. 321–31.

Evans, M. (1980) *Housing, Climate and Comfort.* London: The Architectural Press.

Fanger, P.O. (1970) *Thermal Comfort: Analysis and Applications in Environmental Engineering.* London: McGraw Hill.

Krier, R. (1984) *Urban Space.* London: Academy Editions.

Nieuwolt, S. (1986) Design for Climate in Hot, Humid Cities, in *Urban Climatology and its Applications with Special Regard to Tropical Areas,* WMO-No. 652, pp. 514–34.

Oke, T.R. (1988) Street Design and Canopy Layer Climate. *Energy and Building* 11, pp. 103–13.

Salleh, E. (1994) *Tropical Urban Outdoor Environment and Human Thermal Comfort,* Ph.D. thesis, London: The Architectural Association Graduate School.

Sham, S. (1987) *Urbanization and the Atmospheric Environment in the Low Tropics: Experiences from the Kelang Valley Region Malaysia.* Monograph. Universiti Kebangsaan Malaysia, Bangi.

Spain, S. (1986) The Upper Limit of Human Control from Measured and Calculated PMV-values in a National Bureau of Standards Test House. *ASHRAE Transaction* 92, Pt. 1B.

Terjung, W.H. and Louie, S.S.-F. (1974) A Climatic Model of Urban Energy Budgets. *Geographical Analaysis* 6, pp. 341–67.

Terjung, W.H. and O'Rourke, P.A. (1980) Energy Exchanges in Urban Landscapes – Selected Climatic Models, in *Publications in Climatology,* Vol. XXXIII, no.1, C.W. Thornthwaite Associates/ University of Delaware.

Todhunter, P.E. and Terjung, W.H. (1988) Intercomparison of Three Urban Climatic Models. *Boundary-Layer Meteorology* 42, pp. 181–205.

Part V

EXPERIMENTAL SUSTAINABLE PROJECTS

12 TROPICAL AND TRADITIONAL: INVENTING A NEW HOUSING MODEL FOR THE OLD 36 STREETS QUARTER IN HANOI, VIETNAM

Shoichi Ota

Institute of Industrial Science, University of Tokyo

Abstract

Hanoi's old quarter, also known as the "Old 36 Streets Quarter," composed of extremely long and narrow tube-like houses, is nowadays regarded as a historic district with its traditional way of life. On the other hand, the quarter holds the commercial core of the city as well, and is exposed to the pressures of urban development.

The Quarter is essentially a high-density populated area full of urban and environmental problems. Existing houses in the quarter are already fully used and have been modified ad hoc to meet ever-changing residential needs during last several decades. This shows the gap between structure and usage. The poor infrastructure is insufficient to accommodate the increasing population of the city, and their increasing energy demands such as for air-conditioning.

The Hanoi Experimental Housing Project was started to resolve these difficulties, by introducing a novel housing model based on the study of existing urban fabrics.

The architect Kazuhiro Kojima introduced an important idea, "space block", to the project. Space block is an architectural design methodology, using some basic space blocks (BSBs) as constitutive elements, which form a whole porous structure. This idea is well-matched to the required building performance. BSB is equivalent to housing in a tube-like house. The porous structure creates wind corridors connecting several inner void spaces or courtyards, which encourages natural ventilation and is effective in reducing energy consumption.

The experimental construction was completed in 2003. The model is designed to accommodate local lifestyle and to ensure comfort in the tropical climate. For "sustainable development", we have to consider how to co-exist with conventional urban contexts, in historic and environmental terms. The Hanoi experimental house is not merely a new architectural model, but also takes in old elements of the site and surroundings. This merging of old and new is a key characteristic of the project.

Keywords

Tropical, architecture, sustainability, ecology, Vietnam, shophouse.

12.1 INTRODUCTION

The capital city of Vietnam, Hanoi, has a thousand-year history since it was founded in 1010. Its old quarter, known as the "Old 36 Streets Quarter", is still the commercial centre of the city and provides the core of urban activity.

The Hanoi architectural and urban study team, composed on the Vietnamese side by members from the Hanoi University of Civil Engineering, and on the Japanese side, mainly from the University of Tokyo, has performed continuous research more than ten years. The team has conducted investigations on Hanoi's various aspects from French colonial architecture to the Old Quarter, involving many researchers from various disciplines, such as architectural history, planning and environmental engineering.

Visiting the site so frequently, and staying for a considerable time, the team researched how its inhabitants lived and its complex structure, and the serious problems with which the city is currently confronted.

In 1999, the team started to develop a new housing type based on academic research, aiming to fit in with the local conditions, and supported by Japan Society for the Promotion of Science. When investigating traditional architecture and cities, the various feature of Hanoi's Old Quarter came to light. Indigenous wisdom from the land was applied to solve the problems, and realizing it for Hanoi's future is the task of the team.

12.2 RESEARCH ON HANOI'S OLD QUARTER

12.2.1 Spatial features

The Old Quarter of Hanoi (Figure 12.1) was composed of shop-houses, standing one behind another, having a width from 2.5 to 5.5 m, and a depth of sometimes over 60 m. Shophouses, called "Nha Hinh Ong" or "tube-like-houses", after their narrow proportions, filled the whole of each block (Figure 12.2). In an area where merchandise lines the streets, it was essential that houses face the street to secure shop space. With a façade open to the street and shop space on the ground floor, each individual house

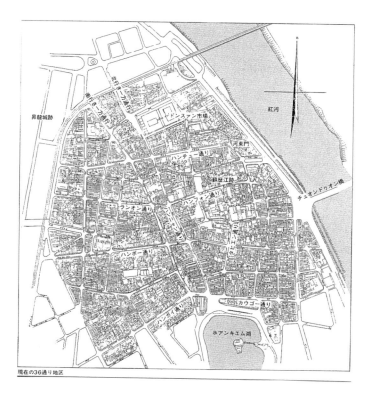

12.1.
Hanoi's Old Quarter. (*Source:* Author)

12.2.
A block in Hanoi's Old Quarter. (*Source:* Author)

had enclosed walls on the other three sides (the back and flanks), so that it had no connection with its next-door neighbours. Thus only on the street could each household communicate with others.

The inner space of shophouses was filled with unique devices not only to preserve the inhabitants' privacy but also to create comfortable living spaces. On the narrow site, a built space and a void space stand one after the other in a rhythmical sequence. Inner courtyards play a role in ventilation and lighting.

The shophouse can accommodate many inhabitants at an average population density of 1000 persons/ha, which is almost the same figure of that of collective housing in the suburbs of Tokyo. They can accommodate so many people only by making the most use of the existing structure.

As to environmental performance, the shophouse can prevent overheating of inner spaces by avoiding direct sunlight, providing ventilation and using materials with sufficient heat capacity.

It has some problems, such as the shortage of floor area due to overpopulation, and loss of ventilation efficiency due to decreased inner courtyard area caused by irregular attached buildings, etc. There is much room to improve its interior environment.

The Quarter was created to introduce a high-density commercial area. With this aim, the mass-volume of shops was calculated and the width of each house was designed based on the scale of shop one household could run by itself. The inhabitants in the area could also bepotential consumers. The depth of the house decides the number of its inhabitants. That number, based on the supposed population of the city, determines the average size of a block, and furthermore, the density in the block, in a ward and the overall spatial structure of the Old 36 Streets Quarter.

12.2.2 Historical background

Originally, Hanoi was founded in 1010 as the seat of Ly dynasty, the first long-lasting Vietnamese feudal dynasty. Along with other East Asian countries, the Vietnamese nation accepted Chinese civilization, from practical aspects such as the governmental system to ideological aspects such as cosmology. According to Chinese ideas, the capital city was supposed to be designed according to a special plan. Thang Long city, the former name of Hanoi, was created around the palace and the city was surrounded by a double wall. The inner city was reserved for royal residences and governmental institutions, while the outer city was for ordinary citizens, and this area was divided into a number of blocks.

In the 14th to 15th Centuries, the outer city was transformed into a market town. Each street was occupied by a specific guild handling particular agricultural or handicraft products. Streets functioned as market places and along them shophouses emerged and were aligned one by one. An alias of the Old Quarter was

"36 Streets", after the number of these guilds in this period (Hung and Thong, 1995).

The spatial structure of the block also changed at that time. Previously, houses on the block stood separately or unconnected with streets. In the later period, shophouses were tightly built along streets and each house was laid out street-wise. The inner part of each block remained undeveloped land and usually had a pound. These vacant areas were, however, gradually occupied as the back parts of each shophouse. Finally, the block was filled up with houses and the shophouse came to have an extremely long and narrow shape.

Through the colonial period shophouses were used as shop space and the residences of wealthy families. After the revolution, the houses were forced to accommodate many others who could not afford to have their own houses or who were appointed to live in the capital city under communist policy. The inner space of shophouses was divided and distributed to each household, and side corridors were created for residents living in the back part to pass through. The spatial structure we can see today in the shophouse was formed in this process.

12.2.3 Urban fabric

Consequent upon its historical growth, the Old Quarter has a peculiar urban organization called "phuong", which has its origin in the Chinese capital city plan, and, in this context, the term means a city block. Despite its original meaning, a phuong in Hanoi does not consist of housing blocks surrounded by streets, but of housing rows on both sides of a street. Here the street is an axis of the phuong, and it forms a framework for the city. The street was put between two rows of houses. The street space was a market and both rows of houses were occupied by a guild associated to a specific occupation and common home village. A street guild has a core institution of a "dinh", a worship and assembly hall enshrining a deity representing their vocation or land. This structure has generally continued in present urban fabrics and provides a base for many sorts of social organizations, such as police, veteran, youth and women's associations, or the communist party. It is also the minimum unit of local government.

The composition of a shophouse reflects this urban fabric. For each shophouse, the façade is like an active skin strongly connected with street market, while its sides and back are just boundaries and have no function in terms of urban fabric.

12.3   REQUIREMENTS FOR NEW HOUSING

What is required of housing in Hanoi's Old Quarter? The Quarter is now suffering from many urban problems.

First of all, shortage of floor space is serious. Under the socialist regime, transformation of shophouses into collective housing made the inner spaces too cramped. Handed down to the younger generation, and usually divided between brothers by inheritance, they are getting narrower. To increase the floor area is the first of the inhabitants' demands for housing. Secondly, there is a problem derived from the first. The response of inhabitants to the insufficient floor area was to use courtyard or rooftop spaces for extensions. Consequently, the interior of the shophouse became tightly packed; and from this follows the next problem. Initially, ventilation and lighting could be through the courtyards; however, this does not now work well due to the irregular extensions. As a result the interior environment deteriorates. Rooms are darker and the heat in the interior space is increased.

The inhabitants of shophouses attempt to solve these problems by using modern equipment, such as electric lighting and air-conditioning, but this caused a problem on the whole city level. Namely, it is energy consuming, and the present urban infrastructures of this city cannot support it. The electricity supply in the summer is becoming overloaded, which forces power cuts. It is essential to reduce energy consumption by making dwellings fitted to the tropical climate.

However, not all new houses will be accepted. Only a housing type fitted to the local context will be adopted. A housing type fitted to Hanoi's Old Quarter will be one adapted to the local urban fabrics. It will be a practical design, that can be constructed within a fixed lot and promote interaction in the street. It will house a shop space facing the street, and have the capacity for collective habitation.

Because of high demand to increase floor area, it will be important to invent a large type of house. Nevertheless, we did not suggest a house far surpassing the present ones in scale. This is crucial for sustainability of the city. It is essential to consider the urban fabric of the city. The Old Quarter still holds a core of commercial businesses and, for this reason, the Quarter is prosperous and thus placed under aggressive development pressure. However, will this be everlasting? The distribution system in the Hanoi urban area is gradually shifting from conventional human labour distributing agricultural products to modern mass-distribution. Market activity in the city is changing, and this will affect the position of the Old Quarter, situated as it is on a conventional network. If this lasts, the economic status of the Quarter will decline relatively. The same tendency can be observed in the whole urban area. Today, the area of the city with an increasing population is suburban zone, with many collective housing estates where the middle class is growing and many people can afford to enjoy city life. They are office workers and bureaucrats, shopping at supermarkets, and having a different origin in a different homeland. A lifestyle distinct from conventional urban fabrics has already emerged here.

For these reasons, it would not be appropriate to assume that the Old Quarter will grow excessively. We have seen many cases when surplus capacity unsuitable to local needs became a burden, and then went out of control. On this occasion the proposed new architectural model is based on the population and present number of households and what is predictable within naturally increasing, architectural design. It was decided to assume an affordable scale of housing.

When constructing new buildings in Hanoi's Old Quarter, we should remember that it is a historic district. The Quarter is designated as a preservation area, and the authorities have enacted a series of regulations for this reason and make efforts to support architectural conservation and regional development. The architectural model should follow this trend. The conservation regulations include slant-line control, prohibiting construction beyond a line looking up from the street, and control of architectural forms to harmonize with traditional architecture. Naturally, it is also required that the new architectural model be connected with tradition. For example, a model with the architectural features of the shophouse, smoothly accepting the traditional way of life. It can be an answer to the conservation of intangible and everyday culture, which Hanoi city promotes. Present architectural conservation seems to stick too much to controlling the exterior, which leads to static preservation in most cases. For conservation of an area full of urban activities, it is necessary to consider urban conservation as dynamic process. Responding to this, the concept of conservation of spatial features is presented in the experimental housing project. Combination of buildings and courtyards, and collective habitation developed through the historical process, are elements of the local character. To maintain this character is a variation of architectural conservation. Retaining a lifestyle in the shophouse dwelling space to accommodate daily activities around the courtyard or in an ambivalent space between interior and exterior is an essential feature of Hanoi. An architectural form to maintain this spatial feature is needed. Only a piece of architecture coexisting with the character of the city and able to respond to the problems of the city can contribute to sustainable development of that city.

12.4 CREATING A NOVEL ARCHITECTURAL MODEL

12.4.1 Design policy

The team engaged in designing a new architectural model for the Quarter with regard to the character of the city. In its basic design, the existing lot was used as it was and no alteration was envisioned, such as connecting with the next-door neighbour or creating a passageway penetrating the lot. Consequently, the planned house only has an opening facing the street, and both sides and the back are enclosed by dividing walls. The façade will interact with the street space, and a shop space will be created also facing the street, aiming to preserve the features of the shophouse.

Inside, the housing will take in the spatial rhythm of repetition of built and void spaces. Inserting courtyards appropriately provides ventilation and lighting, and will be an important tool for passive climate control to reduce energy consumption. In this experimental housing, this spatial composition is three-dimensional, void spaces are connected vertically by "wind corridors". Interior layout is based on conventionally connecting dwelling cells through inner passageways, and each household is distributed from front to back, not stacked on top of each other in storeys.

The issue of conservation of the Old Quarter must be treated seriously. According to the promulgated regulations, the experimental house will be constructed below the height control line and have a slanting roof shape topped with roof tiles to preserve a traditional silhouette. To conform to the preservation of spatial configuration three-dimensional courtyards are used.

12.4.2 New design method

On developing the new architectural model, the architect Kazuhiro Kojima applied his own design method of the "space block" and "porous model".

Space block is a method to design basic volumes combining cubes with a basic measurement. The structure, composed of these basic volumes (BSB: Basic Space Block, Figure 12.3), is called a porous model (Figure 12.4). This model can produce diversified spatial layouts, knitting interior spaces and exterior spaces together. Even in a multi-floor house it responds to contemporary demand for increased available floor area, the model can also provide ventilation and lighting by vertical inner courtyards.

The experimental house was designed using this method on a site in the 36 Streets Quarter. The actual model has multiple floors

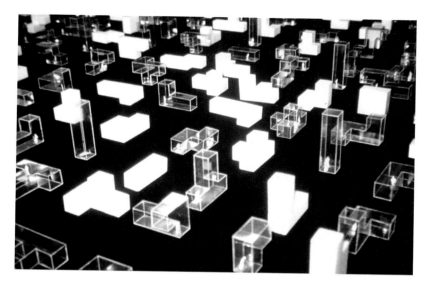

12.3.
Basic Space Block. (*Source:* Kazuhiro Kojima)

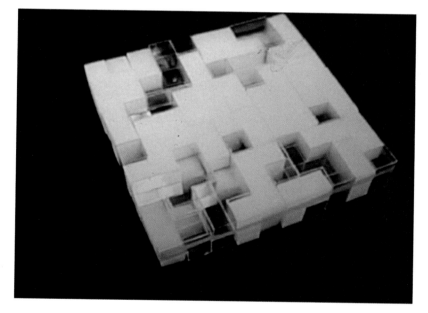

12.4.
Porous structure. (*Source:* Kazuhiro Kojima)

(four floors in each housing unit), which allows the total floor area to increase from 179.9 m^2 in the existing house to 376 m^2.

12.5   ENVIRONMENTAL MEASURES

The biggest theme for our project was to propose a new housing model providing comfort by decreasing environmental loads in a tropical climate. To achieve this, the model employs a passive

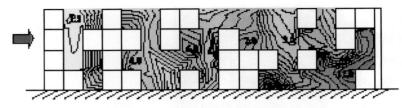

Distribution of air age at the first stage study

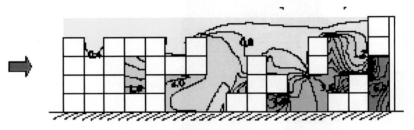

Distribution of air age at the final version

12.5.
Evaluation of ventilation –
distribution of air by age.
(*Source:* Kato et al., 2003)

climate control system. Existing active control by air-conditioning consumes too much energy and is limited by the electric power supply in the area. Moreover, it does not fit with the present Hanoian way of life of spending much time outdoors.

The porous model is designed to increase natural ventilation by creating a wind corridor, placing a radiating cooling panel in the wind corridor, and blocking heat from direct sunlight by using double roofs. Windows and doors on the wall open as high as the ceiling for efficient ventilation, sweeping away the heated and polluted air in the room (Figure 12.5). All of these methods harness natural wind ventilation to reduce air temperature inside. A series of studies was carried out to identify the most effective way to create this wind corridor.

12.6 PROBLEMS WITH THE ACTUAL CONSTRUCTION

12.6.1 Alteration of the design

The Hanoi experimental house was originally intended to be built on the actual site in Hanoi's Old Quarter (Figure 12.6). The team selected the site and negotiated with the residents to allow construction work. However, a serious problem arose, leading to abandonment of this plan. This was because of the too-complicated land ownership of the site. Just one opponent, among those

12.6.
Planned construction site.
(*Source:* Kojima Lab and
Magaribuchi Lab)

claiming ownership, could stop the work. The team failed to get
the agreement of one tenant on the site despite every effort.

The team moved the construction site into the grounds of Hanoi
University of Civil Engineering, located just a few kilometres south
of the Old Quarter (Figure 12.7). Construction work began safely
and soundly through the courtesy and generous understanding of
the rector.

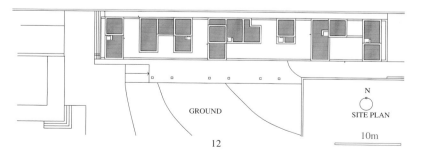

GROUND

N

SITE PLAN

10m

12

12.7.
Actual construction site.
(*Source:* Kojima Lab and
Magaribuchi Lab)

Due to this change, the architectural design of the experimental house was completely altered (Figures 12.8 to 12.13). However, the basic conditions were preserved, in spite of the site change, as the aim of the project was to propose a new housing type for the Hanoi Old Quarter. The construction was assumed to be done on the original site and design work continued on this presumption. Even though some design points had to be changed, the basic concept including the composition of inner spaces (Figures 12.14 and 12.15)

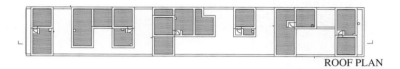

ROOF PLAN

+12.6LEVEL PLAN

12.8.
Plan of the experimental house. (*Source:* Kojima Lab and Magaribuchi Lab)

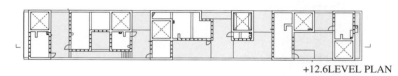

4F PLAN

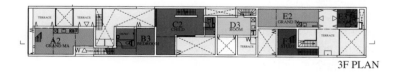

3F PLAN

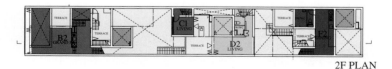

2F PLAN

1F PLAN

10m

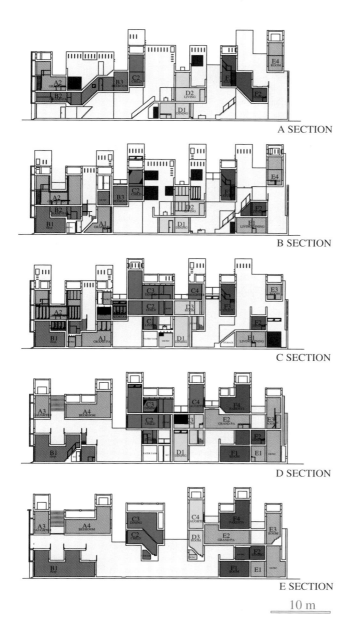

A SECTION

B SECTION

C SECTION

D SECTION

E SECTION

10 m

12.9.
Section of the experimental house. (*Source:* Kojima Lab and Magaribuchi Lab)

was not altered, although the exterior was redesigned, to fit the actual surroundings on the campus. Physically, an abstract form was used to harmonize with the surrounding Modernist buildings. This form was also introduced to reflect the character of the site as virtually a tabula rasa. The final design proposal was like piled-up white cubes.

This alteration showed the possibility of fitting the model to its surroundings, if construction in the Old Quarter is realized then it will be adapted.

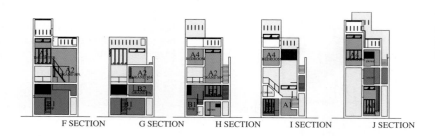

F SECTION   G SECTION   H SECTION   I SECTION   J SECTION

K SECTION   L SECTION   M SECTION   N SECTION   O SECTION

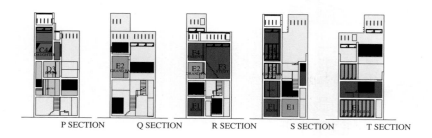

P SECTION   Q SECTION   R SECTION   S SECTION   T SECTION

U SECTION

10 m

12.10.
Section of the experimental house. (*Source:* Kojima Lab and Magaribuchi Lab)

12.6.2 Construction process

Vietnamese construction firms undertook the construction work under the supervision of Japanese architects. All of the construction materials were sourced within Vietnam, except for some unique equipment such as radiating cooling panels.

The experimental house was built using a reinforced concrete structure. At first, frames made from reinforced concrete were

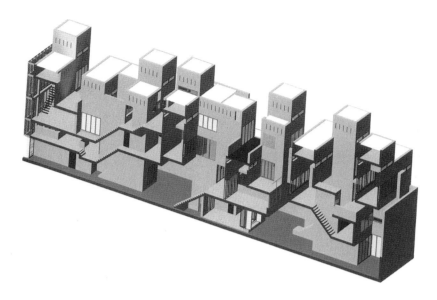

12.11.
Model of the experimental
house. (*Source:* Kato et al., 2003)

12.12.
Experimental house. (*Source:*
Author)

erected, then bricks filled each span of the frame to make a wall. Careful supervision was required, especially in using a cantilever structure to form the vertical courtyard. The completed building was surfaced with mortar and finished with white paint. These construction methods are exactly the same as for ordinary housing construction in Vietnam, which allows relatively low-cost construction.

12.13.
Façade of the experimental house. (*Source:* Author)

Thanks to the efforts of the construction workers, the experimental house was completed at almost the average cost of ordinary house construction in Hanoi, which is important for dissemination of this model.

12.7   COMPLETION AND EVALUATION

The Hanoi experimental house was completed in September 2003. Through the model that the team presented, an architectural example was offered that respected the way of life around an inner courtyard, and able to cope with the urban structure and narrow lots in the Quarter. The model is designed to reflect local open-air lifestyle, to ensure comfort in a tropical environment. Empirical proof of the concept continues.

Modern active climate control, using air-conditioning in enclosed interior spaces, is gradually spreading in Vietnam. From the physical

12.14.
Interior views of the experimental house. (*Source:* Author)

12.14.
Continued

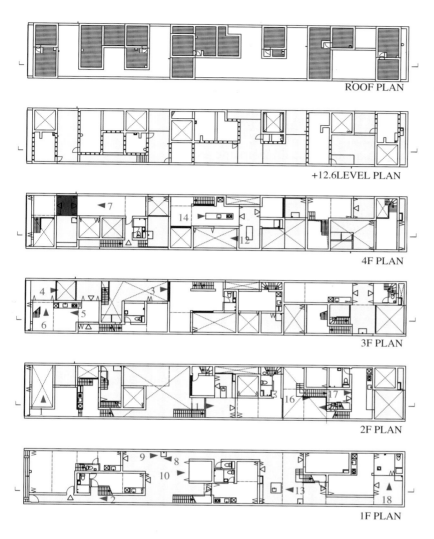

ROOF PLAN

+12.6LEVEL PLAN

4F PLAN

3F PLAN

2F PLAN

1F PLAN

12.15.
Location of the photographs shown in each number corresponds to a photograph and shows the direction of sight.

point of view, the spread of this system will be limited according to the electricity supply or environmental capacity. In a social context, on the other hand, having air-conditioning can be regarded as a success symbol showing the affluence of the owner. Whether our model has a value surpassing this, and how the inhabitants evaluate it, are questions that could determine the destiny of the experimental house.

It is not enough for new housing for tropical regions just to be novel. Whether it is in the tropics or elsewhere, there are local contexts and conventional architectural characteristics to be observed. Only a model responding to and articulating these particularities can take root, otherwise it will not be worth constructing in the real world.

REFERENCES

Kato Shunsuke, Shuzo Murakami, Koichi Takanashi and Hidekuni Magaribuchi (2003) *Hanoi model/Tokyo model – Development of Urban and Building Models for Densely Populated Area with Minimized Environmental Load in Hot and Humid Climate*. Tokyo: Japan Society for the Promotion of Science. (In Japanese.)

Tran Hung, Nguyen Quoc Thong (1995), *Thang Long-Ha Noi Muoi The Ky Do thi hoa*, Hanoi: Nha Xuat ban Xay dung (In Vietnamese).

13 ECOPET 21: AN INNOVATIVE SUSTAINABLE BUILDING SYSTEM FOR ECOLOGICAL COMMUNITIES IN TROPICAL REGIONS

José Roberto García Chávez

Universidad Autónoma Metropolitana
División de Ciencias y Artes para el Diseño
Departamento de Medio Ambiente
Laboratorio de Investigaciones en Arquitectura Bioclimática
San Pablo 180. Colonia Reynosa Tamaulipas. C.P. 02200
México, D.F.

Abstract

During the last fifty years, an inappropriate use of energy and natural resources on our precious planet has been a common pattern, associated with an explosive population growth and an accelerated intensity of industrial activities, based on an irrational exploitation and burning of fossil fuels, and these include intense energy use in cities and buildings, which are responsible for about one half of total energy consumption in the world. All these trends have provoked a severe damage on the planet's ecosystems. Nowadays, it is not only necessary, but urgent to modify these trends and to apply corrective actions. In this work, an innovative sustainable construction system called ECOPET 21, has been applied in a house prototype and integrated with the application of bioclimatic design principles and sustainable technologies as well as environmental planning, aimed at the development of ecological sustainable communities, particularly for tropical regions. The principles and benefits of this approach are presented and their benefits demonstrated. The main objective of this project is to implement ECOPET 21 with the integration of sustainable technologies into bioclimatic architectural design and rural planning actions in tropical regions, based on the use of renewable energies, taking into account and emphasizing the economic and social issues, aimed at promoting a multiple effect and a consistent sustainable development with local and global benefits.

Keywords

Sustainability, ecological communities, ECOPET 21, design strategies, housing, bioclimatic architecture, climate, comfort, tropical countries, renewable energies.

13.1 INTRODUCTION: ENERGY USE PATTERNS AND ENVIRONMENTAL IMPACT – POPULATION GROWTH AND INDUSTRIALIZATION

13.1.1 World energy use patterns, population growth and environmental impact

With the advent of the *Industrial Revolution*, during the middle of the eighteenth century, the use of renewable energy resources – used from the first traces of human beings – such as firewood, sunlight, wind, waterfalls, among others, began to be replaced. In fact, coal, as the first of these fossil fuels exploited, formed the basis of the so-called *Industrial Revolution*. These energies started to be used and their consumption patterns increased gradually as the population boosted. Most production and anthropogenic activities from the *Industrial Revolution* have been based on the use and exploitation of fossil fuels, i.e. coal, oil and natural gas.

The burning of fossil fuels in power plants, to generate diverse forms of energy, started to provoke the emission of a great number of pollutants to the environment. The rate of this consumption patterns became greater and the build-up of contaminants started to affect the ecosystems in our fragile blue planet. However, the number of inhabitants of the planet at the time around the *Industrial Revolution*: 1000 million inhabitants, was not relatively significant to alter the environment at a large scale. However, since then, the relationship between population growth and energy consumption has been growing exponentially and dramatically. During the *Industrial Revolution*, energy consumption was 10 *million tonnes of oil equivalent/year* (mtoe/year); 100 years later, in 1900, population reached about 1700 million people, with a consumption of 800 mtoe/year. In 1970, population was 3600 million people, who consumed 5200 mtoe/year (WRI, 2002; WEC, 1997). During the 1950s and 60s world energy consumption increased at a staggering rate.

According to recent studies (WRI, 2002), world energy production grew 52% in last two decades of the twentieth century. In 2006, world population is estimated to reach about 6700 million with an energy consumption of more than 10 700 mtoe/year (IEA, 2005; BP, 2005; IPC, 2001). Besides, more than 90% of the global primary energy production comes from the so-called fossil fuels (IEA, 2005), which in turn are responsible for the great majority of the severe environmental damage provoked on our natural habitat. Emissions from combustion of coal, oil and natural gas and from deforestation and land cultivation have increased the natural CO_2 concentration by more than 30% over the past 200 years (31% since 1750), and continues to increase (IPCC, 2001). The pre-1750 CO_2 concentration was 280 ppm, and, according to the recent information provided by IPCC (The Intergovernmental Panel on Climate Change, formed and sponsored by the United Nations), the current tropospheric concentration updated to February 2005 was 375 ppm (IPCC, 2001, updated). This report also indicates that the global average surface temperature has increased since 1861. Over the last century, the increase was $0.6 \pm 0.2°C$. The globally averaged surface temperature, based on a number of climate models, is projected to increase by 1.4–5.8 K over the period 1990–2100 (IPCC, 2001). The atmospheric concentration of CO_2 is currently rising at about 0.5% per year with annual emissions of about 7.2 GtC (gigatons of carbon. 1 GtC equals 1000 million metric tons) – that is at about a rate of 1.0 CO_2 tonnes/person/year, from fossil fuels combustion and land-clearing activities.

Therefore, according to the recent IPCC findings, there is stronger evidence that most of the warming and the consequent climate change, observed over the last 50 years, are attributable to human activities. Unfortunately, should this situation persist, anthropogenic influences mainly due to irrational energy use patterns, population growth, intensive industrialization and urbanization, will continue to change atmospheric composition throughout the twenty-first century. If these changes take place, they would provoke global catastrophic consequences.

13.1.2 Application of renewable energies and sustainable technologies to promote the development of ecological communities in tropical regions

In 2006, the world will reach a population of 6700 million, which will require more than 10 700 million tons of oil equivalent, and 90% of this energy comes from fossil fuels, with the multiple consequences this situation provokes on our precious blue planet (IEA, 2005; IPC, 2001). On the other hand, renewable energies, such as solar, wind, wave, and geothermal, biomass, among others, have a significant potential and countless advantages in their application and use for all types of energy needs of the inhabitants of the planet. The global potential of natural renewable energies is enormous. For instance, with the tiny part of the solar energy that reaches the Earth surface, all worlds' energy demands could be solved 20 000 times (Rostvik, 1992). Other renewable energy sources, such as wind, ocean and so on, have also a large potential for covering energy needs and are promising alternatives for solving most environmental damage.

13.2 THE HOUSING PROBLEM AND ITS RELATIONSHIP WITH THE ENVIRONMENT IN TROPICAL REGIONS

If the world's current energy consumption is more than 10 700 mtoe per year, the building materials process (including embedded energy), construction, operation and maintenance of buildings represent about half of this. This means that buildings are responsible for a significant amount of the environmental damage provoked on our planet and therefore, any sustainable action carried out in communities, both urban and rural, will imply a benefit for our natural habitat. This work focuses in the premise that buildings and communities offer great opportunities for contributing to reduce energy consumption and then environmental damage. The objective of this chapter is to present a number of innovative sustainable strategies which can be applicable in both urban and rural communities, particularly in tropical regions.

13.2.1 Provision of housing: application of sustainable strategies for rural communities located in typical tropical regions

Provision of an adequate shelter has been a basic premise that our ancestors sought to accomplish since remote times. Certainly, having access to a safe and healthy shelter is essential to a person's physical, psychological, social and economic well-being and it should be a fundamental part of international actions. Therefore, supplementary conditions related to adequate housing must include clean drinking water, energy for cooking, heating and lighting, sanitation and washing facilities, food storage facilities, refuse disposal, site drainage and emergency services, among other facilities. However, all these services are rarely available in most tropical locations of the world, particularly those in developing countries.

The human right to housing has been explicitly set out in the *Universal Declaration of Human Rights*, as well as in other international human rights treaties and declarations. The *Universal Declaration of Human Rights* states precisely that:

"Everyone has the right to a standard of living adequate for the health and well-being of himself and his family, including food, clothing, *housing* and medical care and necessary social services, and the right to security in the event of unemployment, sickness, disability, widowhood, old age or other lack of livelihood in circumstances beyond his control." (UN, 1948).

Recent calculations by the UN Centre for Human Settlements estimate that over one billion people worldwide live in inadequate housing and 100 million are homeless (HABITAT Agenda, 2002). According to other figures, released by the World Health Organization, and in the Johannesburg World Summit on Sustainable Development, 1500 million people in developing countries do not have access to potable and safe drinking water and 2500 million people live without access to adequate sanitation services; and 3.4 million people die every year from water related diseases (Johannesburg Rio + 10, 2002; WHO, 2004). These figures indicate that there is an urgent need for actions to reduce the adverse effects and to promote alternative strategies based on sustainability at global and local levels.

13.2.2   Application of sustainable strategies in buildings to achieve comfort conditions – bioclimatic design

One of the most important roles of a building is to provide shelter from the adverse surrounding factors. This was certainly the fundamental premise of ancient and traditional architecture. Unlike traditional architecture that consciously responds to climate, culture and traditions, most modern architectural examples ignore these factors and use energy and natural resources in a way that is far from the present sustainable development principles. By means of natural passive, energy-efficient building design, that is bioclimatic architecture, associated with the utilisation of local and low embodied building materials, the requirements for achieving ambient comfort conditions for occupants can be met in most climatic regions worldwide, including tropical regions.

The main objective of a common sense *bioclimatic architecture is to provide maximum ambient comfort conditions at minimum expenditure of energy, while protecting the natural environment of the planet.* This approach includes the achievement of holistic ambient comfort conditions, which meet the hygrothermal, luminous/visual, acoustic, olfactory and air quality requirements of the occupants.

However, the great majority of the contemporary buildings around the world are still built under the so-called *International* and *Post-Modern Architectural Styles*. These buildings are characterised by a high dependence on artificial ambient control systems and large consumption of fossil fuels, which are some of the main causes of the existing environmental damage on the planet. These artificial systems are used in buildings to provide heating, cooling, ventilation, lighting and also used for cooking and to heat water.

This situation has provoked a number of health, environmental, economic and even social and cultural problems. All of those requirements can be satisfied to some extent from the application of *bioclimatic architecture, and the use of the diverse renewable energy resources available in the world*, particularly in tropical regions. These actions are also aimed at reducing the dependence on energy consumptive services and the emission of pollutants to the atmosphere, whilst improving the quality of life and the natural environment.

13.3   INTEGRATION OF SUSTAINABLE STRATEGIES: A CASE STUDY IN A RURAL COMMUNITY LOCATED IN A TROPICAL REGION

13.3.1   Principles of the application of bioclimatic design and sustainable strategies in buildings

The principles of the application of sustainable strategies in a rural community proposed in this work are aimed at improving the people's quality of living, based on the integration of bioclimatic design with sustainable technologies, taking advantage of the great potential of renewable energy resources, so that practitioners, policy decision-makers and researchers, through the results and experiences of this project, might have a useful source of information for the application of sustainable development actions. This eventually contributes to generate a multiple effect in other tropical locations around the world, with the objective to improve and preserve our natural environment and the quality of living. Besides, this project also has social, economical, educational and health benefits, as it can contribute to promote people to settle down in their local communities, which is an essential factor to prevent that large urban centres continue to grow at currently expanding rates and to promote a more balanced situation in urban and rural communities. Therefore, this project is based on the application of the following strategies:

- Integration of bioclimatic design by means of the application of passive cooling and heating systems for providing natural climatization for achieving comfort conditions of the occupants, whilst reducing the energy consumption in buildings.
- Construction of houses of the community by supervised self-construction methods. Women and children are the main actors of this process.

- Houses of the community are integrated with sustainable technology systems, to get a comfortable, healthy and sustainable habitat. Regional construction materials and recycling materials are used. Alternative and solar energy systems integrated in the houses include: solar collectors for water heating, solar cookers, solar refrigerators, solar distillers, solar driers, PV systems, and firewood high thermal efficiency saving stoves, among others.

- Development of a new *natural resources culture*, based on the implementation of alternative and solar energy systems for production in the houses, oriented to energy self-sufficiency. People of the community are trained in the construction, operation and maintenance of their houses and their solar energy and ecological systems.

- In terms of water, a new *water use culture* has been promoted among local people of the community, based on the training to construct, operate and maintain a rainwater collection and storage system, a greywater recycling system, a rainwater absorption well to stabilze ground water table; and low consumption and water use efficient devices. At present, most regions of the world confront a severe water scarcity problem and this work is oriented to provide alternatives and corrective measures for solving it.

- As to waste-treatment, a new culture on *waste treatment* has been encouraged in the community to separate and classify organic and inorganic wastes, to make compost and produce biogas. Direct benefits for the community have been promoted through the implementation of recycling material storage centres.

- In terms of life-supporting systems, the community inhabitants are trained to integrate, construct, operate and maintain in their houses a family vegetable garden; an orchard; an aquaculture pond with trout, tilapia and carp; a digester, and a farm area with chickens, rabbits and bees, among other systems.

- As to improving the economic and social conditions of the community, people are prepared and trained to implement and operate *productive projects*, through community workshops and co-operatives, agricultural and flower growing industries, as well as family micro enterprises, among other activities, based on the traditional skills of local people.

As a representative example of the application of the principles and sustainable strategies mentioned above, it has been proposed the development of an ecological sustainable community project, recently developed in a typical rural location in Mexico.

13.3.2   Location of Mexico: climate conditions and availability of energy resources

Mexico is a country located within a tropical region in the northern hemisphere, with geographical coordinates: 14° 31′ to 32° 43′ North Latitude, and from 86° 42′ to 118° 22′ West Longitude. The country has 2 million km^2 of land surface and a total population of 106 million people, estimated for 2006 (CONAPO, 2002). Mexico is located in a privileged region with a great diversity and availability of energy and other natural resources formed since remote times. In terms of energy from conventional fossil fuels, such as oil, coal and gas, these sources are abundant in the country. Recent figures indicate that hydrocarbons are the main source of primary energy production, accounting for 90% of total (SE, 2005), and this makes Mexico a mono-dependent country, with its primary energy coming mainly from fossil fuels. In terms of renewable energy sources (RES), such as solar, wind and so on, Mexico is also located in a privileged geographical region with plentiful of them.

13.3.2.1   *Situation of the lack of housing in Mexico*

The housing deficit in Mexico is more evident in areas where people live under the line of extreme poverty conditions, which at present represents more than 50 million people. Most of these regions are located in tropical areas. More than 150 000 communities in rural areas, remote or isolated, as well as suburbs surrounding main urban centres are representative of this situation. Rural communities, representing 15% of total population, are mainly found in regions where 80% of national hydro-energy resources are located. This is a favourable condition that could contribute to solve some of the problems these communities are facing. Therefore, the application of renewable energy resources is a key factor which can promote the necessary changes to diminish and revert the current environmental damage.

  This deficit of houses in Mexico is mainly due to demographic growth, uneven geographical population distribution, lack of land planning, land use speculation, lack of self-construction capabilities, lack of suitable traditional construction knowledge and high cost of conventional building materials and labour, among other factors. The current situation in Mexico indicates that there is a severe lack of housing, with a deficit of about 6 million houses. This figure indicates that just to meet the demand, the annual construction rate needs to be of about 750 000 houses until the year 2025 (SEHB, 2003). Nonetheless, according to NGO's sources, the real deficit is higher (CANADEVI, 2003). Approximately two-thirds of housing production is self-production, without finance, in irregular settlements, with high construction costs and lacking basic service facilities.

13.3.2.2 *Factors affecting availability and affordability of housing in Mexico: use of typical building materials and cost of land*

The lack of housing in the majority of urban and rural areas of Mexico and its relative high cost provokes social conflicts, affecting people's health and preventing families to be suitably integrated. Inadequate housing in urban areas is the main factor that causes an irregular land occupation, as well as frequent moving of people from out-of-control urban locations to high risk zones or environmental protected natural areas, and this in turn results in a burden for the local authorities as they have to pay very high costs for urbanization and provision of basic services, and very often, this provokes irreversible ecological damage. Along with the expectations for higher salaries, the hope for getting a better house and the search for better opportunities are some of the strongest motivations for people to migrate to large urban centres or very frequently, outside the country too. This situation is not exclusive of Mexico and is very common in other developing countries located in tropical regions around the world.

As a common practice for building houses, the use of industrialised materials has become very popular in Mexico, while the use of traditional and indigenous materials has been declining and undervaluated. The cost of land has also increased at very high rates due to speculative commercial interests. This situation has provoked a huge increase in the cost of houses and this in turn prevents the potential users to afford this basic infrastructure. Neither the public nor the private institutions can build the vast number of houses needed to satisfy the demand. Besides, the users have lost the ability for *self-construction* of their own houses in most regions of the country. It is precisely through a well planned and *guided self-construction method*, proposed in this research work, that the housing demand can be diminished.

Regarding availability, apart from a large stock of conventional building materials, Mexico has large quantities of indigenous or traditional construction materials, such as earth, adobe, stone, bamboo, thatch, etc., which are available in most locations. However, the use of these materials has been declined due to the loss of capability using traditional construction methods in their application by local builders. In most cases traditional materials, proven to be

13.1.
Typical tropical Mayan house with previous thatch roof, substituted by a more "modern" galvanised material. (*Source:* Author)

very effective for providing comfortable conditions in extreme climatic conditions of traditional housing, have been superseded by new materials (Figure 13.1), which in turn have affected both the inner ambient conditions of the occupants and caused an increase of energy consumption.

Two of the main factors which also prevent people to have access to domestic dwellings are the *high cost of building materials and the high cost of labour.* One alternative solution to this situation is an approach in three areas:

- the use of local low-cost building materials,
- the use of low-cost recycling materials, and
- the application of simple self-construction methods through technical training of local people.

The application of sustainable actions in these three areas has the potential to reduce the cost of housing for both building materials and labour, and this in turn can also promote an increase in the rate of housing which can eventually reduce the high deficit in the country, among other benefits.

13.3.2.3 *Case study house of the ecological community: Innovative construction system of the envelope. Use of recycling materials. Sustainable approach. Use of PET as basic building material*

One alternative to meet the housing demand in Mexico is the use of *low-cost materials and self-construction methods*, associated with appropriate training directed to local people. The use of adobe as a local material for reducing cost is an important alternative as long as it is combined with a suitable training programme and particularly implemented with structural capabilities for use in high seismic areas in the country. Other approach is the use of *recycling materials*, such as those consisted of empty plastic bottles, widely used in the country, and technically known as PET (polyethylene terephalate).

Previous studies have shown that the use of PET, as a basic building material is a promising approach that can be used to reduce construction costs as well as to provide other benefits (García Chávez, 2002a,b, 2004). PET is an ideal container material, widely used for carbonated drinks and other liquids. Because PET is inert and of endless durability, it remains unchanged for centuries (From 100 to 1000 years). These features make PET a suitable material for recycling. However, the common practice worldwide is simply to throw the empty bottles away after use.

According to recent information, Mexico is the second world consumer, after the USA, of soda beverage which uses mainly plastic PET bottles as a container. Every person in Mexico consumes 150 litres of soda beverage per year, which at national level represents 15×10^9 units of PET bottles of 1 litre or 30×10^9 PET bottles of 500 ml (Beverage Digest, 2003). Apart from this high consumption of PET, inhabitants of Mexico consume large amounts of bottled water, containers of which are also made of PET, and that can also be used for practical purposes. Recent figures indicate that Mexico is the largest market of bottled water in the world (Universoe, 2005). The annual consumption of this product is 15 462 millions of litres, being 13 678 millions of litres in big containers of about 20 l each and 1784 millions of litres correspond to PET bottles of up to 3.3 l. Taking into account these figures, there is a large amount of useful material which should otherwise be better used for recycling purposes.

As related to construction using innovative materials, previous studies have shown that PET is a promising alternative to conventional building materials for both external and internal walls and for roofs, provided it is based on a well designed building and modular design system (García Chávez, 2002a,b, 2004).

The main advantages of this innovative building and recycling alternative compared to conventional building materials are: good structural capabilities, lower cost, good thermal and weather proof performance and easier construction and maintenance work. There is also an evident environmental benefit in using PET material for buildings as the dumping of this to the environment can be significantly reduced.

13.3.2.4 *Principles of the innovative construction method using recycling PET as a basic building material of the case study building*

The construction method of this case study building consisted of: walls, made of PET bottles, with a modular panel of 240 cm height × 120 cm width, × 10 cm thick (Figure 13.2). This particular ecological construction system has been called ECOPET 21 (Figure 13.3). The house was built by local people, trained in supervised self-construction. The typical module PET panel was used as the basic reference and sub-modules from these parametric dimensions were also built according to the needs of the project design and aimed at optimising the construction process.

Every square meter of modular panel requires 50 bottles, and the total average number needed for a 90 m^2 house prototype is from 9000 to 12 000 units. Taking into account the total annual production of empty bottles in Mexico, about 1 800 000 houses can be built with this innovative building system every year. This represents a high potential house building to favourably meet with the current deficit.

This project has been the result of a coordinated programme being carried out in the community and is based on training people on how to build their houses through *supervised self-construction*. The principle behind this programme is that a selected team of local people is trained on how to build their own house and once they are skilled in the building process, they become the new trainers of

13.2.
Construction process of PET modules. Perimeter steel frame and wire enclose of bottles. (*Source:* Author)

13.3.
ECOPET 21 construction system,
shown on walls and roof.
(*Source:* Author)

the following trainee's team to build their house and this generates
a favourable multiple effect within a virtuous circle.

In this project, apart from applying bioclimatic design based on
passive solar techniques and using the building envelope for providing thermal comfort to the occupants, daylighting strategies
integrated with energy efficient electric equipments are also implemented. Solar energy and sustainable technologies are integrated in
the house prototypes. For water use, it included a rainwater collection and a storage system, an ecological sewage treatment plant, a
grey water recycling system, a rainwater absorption well to stabilise
ground water table; and the use of low consumption and water use
efficient devices.

Water use is another issue that is essential for sustainability. In
Mexico, there is scarcity of water and at the same time this valuable
natural resource is squandered. As an example, the particular case
of Mexico City Metropolitan Area (MCMA), for 18 million people living in, there is a water supply of 65 m^3/s (70% coming from more

than 1500 local groundwater wells, provoking also a soil sinking problem – an average of 10 cm/year – which in turn intensifies the seismic movements due to the resonance effect, frequently occurring in the MCMA, such as the earthquakes of 1985). This means that the consumption per capita is of more than 360 l/person/day, which is more than double the average water consumption of international standards of 150 l/person/day. Only 5% (3.3 m^3/second) of those 65 m^3/second is treated and used for non-potable services. As a paradoxical situation, the MCMA has a great potential for water supply from rainfall, which at present is regrettably ignored and not used. Annual precipitation is about 900 mm, which represents a tremendous potential for water use, as for each roof area of 100 m^2, 90 m^3 of rainwater can be collected annually.

However, the rapid evacuation of rainfall, due to the existing local sewage systems avoids the collection and reuse of this valuable natural resource, provoking alteration of the hygrothermal balance conditions in the MCMA. This absurd situation is a paradox in the fact that most places face a dual problem: water scarcity and flooding. This situation clearly denotes the problem of lack of water control management and conscious use, associated with a lack of a realistic culture on water use. In MCMA, a precipitation of 900 mm/year deserves to be taken advantage and, taken into account the 5 million houses, with 100 m^2 of roof area, representing 500 million m^2 of potential collection area, and the 900 mm of precipitation, a total of 4.5×10^{10} m^3 which can be collected, and this means that the vast majority of water requirements of MCMA inhabitants can be solved and this in turn implies not only the economical but also the environmental and quality of living benefits.

Therefore, as to water as a basic resource, in the ecological prototype of the community, with an annual precipitation of 600 mm, the implementation of a new water use culture has been proposed, based on the training to construct, operate and maintain a 90 m^2 rainwater collection and a 6 m^3 storage systems, a grey water recycling system, a rainwater absorption well, and low consumption and water efficient use devices, and design and construction of integral treatment systems for water, resources and nutrients.

Waste treatment and life/support systems are also implemented in the ecological community. The final appearance of the prototype house after completion is shown in Figures 13.4 and 13.5. The prototype house built in this community, was based on the application of design strategies for promoting sustainability in the form of a holistic application using innovative building materials, integrated with bioclimatic design principles and sustainable technologies (Figure 13.6). The objectives are aimed at reducing

13.4.
Main façade of the prototype house. (*Source:* Author)

13.5.
North façade of the house prototype. (*Source:* Author)

construction costs while providing suitable indoor thermal and luminous comfort for occupants and high levels of self-sufficiency in energy and water, as well as suitable sewage, waste treatment and food production systems. Results have shown that the application of innovative building materials and construction systems integrated with bioclimatic design techniques and sustainable technologies is

13.6.
General view from southeast showing solar energy and sustainable technology systems. (*Source:* Author)

a promising alternative to reduce housing costs while providing suitable indoor thermal and luminous comfort conditions for occupants, and improving their economy and quality of living as well as the environmental conditions of the region. This approach can also be applied to promote beneficial multiple effects, and if applied at massive levels, it can contribute to reduce the housing deficit while reducing the severe environmental damage, meant effectively to promote sustainability for the existing and new generations of the new millennium at regional, national and global levels.

13.4 DESIGN CONSIDERATIONS OF CASE STUDY: A TYPICAL RURAL HOUSE PROTOTYPE LOCATED IN AN ECOLOGICAL COMMUNITY

In this work, the potential of recycling PET as innovative building material for reducing construction costs while improving thermal comfort conditions has been investigated in a representative case study, as the first housing prototype of an ecological sustainable community located 90 km north of Mexico City. This prototype house is located in a rural location with a climate characterised by the large daily and seasonal temperature swings. The house prototype area is 90 m^2 (Figure 13.7).

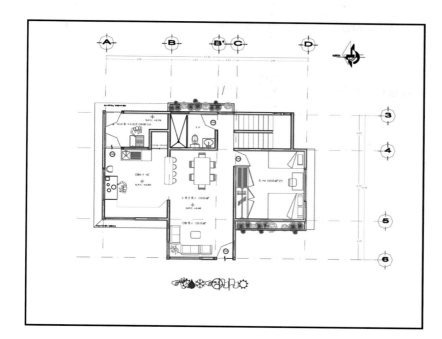

13.7.
Floor plan of prototype house.
(*Source:* Author)

The design of the house is based on a modular grid of 90 cm × 90 cm and applying modules of different dimensions for the various prototype components. The construction method of the house consisted of walls, made of recycling PET panels, 240 cm height × 120 cm width, and × 10 cm thick, as the basic module. For the roof, the same construction system was used, but with double thickness – 20 cm.

The PET panels were built based on this modular design principle, using two layers of sand-mortar plaster 2.5 cm thick on each side of the walls, to make a 15 cm total wall thickness. The roof was built of the same PET system using on top a 5-cm concrete upper compression layer with a 10 × 10 metallic grid in between, to make a total roof thickness of 25 cm.

Under the *supervised self-construction* approach of this project, the main actors of the programme are the local women and children who participate very actively in the building process, particularly in the elaboration of the PET panels (Figure 13.8). Some sequences of the construction process and the energy sustainable systems integrated in the house are shown in Figures 13.9 and 13.10. In this project, apart from applying bioclimatic design based on passive solar techniques for providing thermal comfort to the occupants, daylighting strategies integrated with energy efficient electric

13.8.
PET panels construction process.
(*Source:* Author)

13.9.
Detail of main entrance. (*Source:* Author)

13.10.
Interior view of the prototype house of the sustainable ecological community. (*Source:* Author)

equipments have also been implemented (Figure 13.10). The aesthetics of the exterior and the interior of the house, have also been considered, and integrated with harmonic proportions in its modular conceptualization to come up with suitable results in terms of design (Figures 13.9 and 13.10).

The solar energy and sustainable technologies integrated in the house included a 1 kW PVs system, a 200-l solar collector for water heating, a solar cooker, a solar refrigerator, a solar distiller, a solar drier and a firewood high-thermal efficiency saving stove.

For water use, the house included a 90-m^2 rainwater collection area and a 6-m^3 storage system, an ecological sewage treatment plant, a grey water recycling system, a rainwater absorption well to stabilize ground water table; and the use of low consumption and water use efficient devices.

As to waste treatment, a new culture has been implemented in the case study prototype, based on a sustainable recycling programme.

The prototype house also integrates a family vegetable garden, an orchard, and can also include an aquaculture pond with trout, tilapia and carp; a digester, and a farm area with chickens, rabbits and bees as supplementary food production and life-support systems.

The inhabitants of the community are also trained on how to construct, operate and maintain these systems in their houses.

13.4.1 Economic benefits of the prototype house construction

The construction of the prototype house presented a 40% reduction of total cost. The breakdown of the different cost components showed that by using local labour to build the panels, reduction was 60% relative to conventional brick or concrete block walls. Roof cost was 30% lower than conventional reinforced concrete roof. Relative to a conventional construction and due to the reduction in the cost of building materials and labour, the unitary cost was reduced from 150 US $/m^2–90 US $/m^2, without taking into account the energy and sustainable systems implemented, which when included, increased the cost of the house to 300 US $/m^2. In practical terms and because of the reduction in the final cost, the payback period of the prototype house with the sustainable technologies implemented was 3 years, relative to a conventional house in the region of similar surface.

Therefore, the use of innovative building materials is an effective alternative to reduce construction costs of typical houses located in rural locations and this can also contribute to reduce the current house deficit of a location. In terms of time of realization of the construction, there was a 30% reduction in the case study house, compared to a conventional house of similar dimensions. It is expected that this time can even be shorter, that is more favourable, when more expertise in the house construction of this type is achieved by local builders of the community. Furthermore, the trained local people can also have this activity as a promising alternative to have access to job opportunities and productive projects in the community, aimed to reduce migration and to assist in providing economical means to mitigate the extreme poverty conditions of their families.

13.4.2 Social benefits for implementation of the prototype house

Apart from the promising approach of the application of innovative building systems using low cost materials such as ECOPET 21 for achieving occupants' comfortable conditions in typical buildings of rural communities, there is an additional benefit regarding the contribution to reduce the house deficit in the country. Therefore, this can be a key factor to promote an increase of house construction. Other benefits include the developing of a promising platform for creating jobs and productive projects in poor rural regions, as well as forming the basis for generating building examples which can be used to promote a multiple and positive effect, particularly in tropical regions. Other benefits of this project are oriented to promote social interaction and groups integration for developing activities to improve the quality of living.

13.4.3 Environmental benefits of the prototype house

Furthermore, the use of this holistic approach also has environmental benefits such as the resulting reduction of environmental damage. The preservation and improvement of the environment is a key issue, not only in urban, but in rural areas, particularly in tropical regions.

13.5 CONCLUSIONS

The results of this work have shown that it is possible to achieve a great number of benefits with the integration of sustainable strategies for ecological communities, particularly in tropical regions. As to the implementation of sustainable actions, it has been demonstrated that the use of innovative building materials and construction systems can effectively reduce construction costs of typical dwellings in low income housing and this can also contribute to reduce the housing deficit while providing suitable indoor thermal comfort for occupants, and improving their economy and quality of living as well as the regional environmental conditions. Furthermore, both, the local trainers and trainees can also have this activity as a promising alternative to have access and generate job opportunities and *productive projects in developing communities,* aimed at reducing migration and to assist in providing economical means to mitigate the existing poverty conditions prevailing in most of these locations around the world.

It is expected that the approach of this project can also serve as a demonstrative example and thus be applied to promote beneficial multiple cascade effects, and if handled at massive levels, it cannot only contribute to reduce the housing deficit but also to reduce the severe environmental damage, among other direct and indirect benefits. Therefore, the sustainable design strategies, and techniques applied in this project, based on the integration of bioclimatic design with sustainable technologies and on the use of simple and recycling low-cost local materials, are aimed at helping primarily the poorest local people of tropical regions. This attitude of justice is also a promising environmental, economic and social approach which can also contribute to improve the conditions of our natural environment as well as people's economy, health and quality of living, to promote a holistic and a new approach to evolve towards a more humane and sustainable global society.

REFERENCES

Beverage Digest USA (2003) Website: http://www.beverage-digest.com

BP (2005) Statistical Review of World Energy, June 14th, 2005.

Brundtland, H. (1989) How to Secure Our Common Future, *Scientific American* (Special Issue), 261, no. 3, p. 190.

CANADEVI (2003) *XII Conferencia Nacional de Vivienda*. México.

CONAPO (2002) *Estimaciones de CONAPO, con Base en Proyecciones de la Población de México, 2000–2030*, México.

García Chávez, J.R. (2002a) Innovative Building Materials for Sustainable Applications in Low Cost Housing in *International Solar Renewable Energy Education International Conference Proceedings*. ISREE: Orlando, Florida, USA.

García Chávez, J.R. (2002b) Potential of Innovative Building Materials for Reducing Construction Costs Whilst Improving Thermal Comfort Conditions. PLEA 2002 Proceedings in *Passive and Low Energy Architecture International Conference*, Toulouse, France.

García Chávez, J.R. (2004) Design Strategies for Promoting Sustainability. INTA Proceedings. International Network for Tropical Architecture in *First International Tropical Architecture Conference*. Singapore, February.

García Chávez, J.R. and Fuentes, V. (2001) *Application of Sustainable Development Criteria in a Large Urban Centre*. Published in PLEA, Passive and Low Energy Architecture, vol. 2, pp. 1075–1082.

Intergovernmental Panel on Climate Change (IPCC), United Nations Environment Programme (UNEP) and the World Meteorological Organization (WMO) (1997) *IPCC Second Assessment Report of the Intergovernmental Panel on Climate Change: Contributions of IPCC Working Groups*, January. New York: Cambridge University Press.

Intergovernmental Panel on Climate Change (IPCC), United Nations Environment Programme (UNEP) and the World Meteorological Organization (WMO) (2001) *IPCC Third Assessment Report of the Intergovernmental Panel on Climate Change: Contributions of IPCC Working Groups*, January. New York: Cambridge University Press.

International Programs Center, U.S. Bureau of the Census, Washington, D.C. (2001) World Population. Website: http://www.census.gov/cgi-bin/ipc/popclockw.

IPCC (2001) Climate Change, in Houghton, J.T., Ding, Y., Griggs, D.J., Noguer, M., van der Linden, P.J., Dai, X., Maskell, K.

and Johnson, C.A. (eds), *The Scientific Basis*. Cambridge, UK: Cambridge University Press, pp. 881.

Johannesburg Rio + 10 (2002) *UN World Summit on Sustainable Development. Johannesburg Rio+10 2002*, Johannesburg.

Rostvik, H. (1992) *The Sunshine Revolution*. Stavanger: SUN-LAB Publishers.

Secretaría de Energía (2005) *Balance Nacional de Energía 2004*. SE, Mexico City, December, 2005.

SEHB (2003) Social Enterprise Newsletter. Harvard Business School, Fall 2003. Website: http://www.hbs.edu/socialenterprise/newsletter/.

UN-HABITAT (2002) United Nations Human Settlements Programme. Habitat Agenda 2002 in *World Summit on Sustainable Development*, Nairobi, Johannesburg.

United Nations General Assembly Resolution (1948) The Universal Declaration of Human Rights (1948), adopted and proclaimed by 217 A (III) of 10 December 1948. Article 25.1.

UN-Rio, Agenda 21 (1992) Documento resulltante de la Reunión Mundial sobre Medio Ambiente y Desarrollo Sostenido. ONU, Rio de Janeiro, Brasil, Junio.

UN (1997) United Nations Framework Convention on Climate Change. The Kyoto Protocol, Kyoto, Japan.

Universoe (2005) Website: http://www.universoe.com/ciencia/articulo/ecologia/46_agua.shtml

WHO (2004) Website: http://www.who.int/en/

World Resources (2002–2004) World Resources Institute in collaboration with the United Nations Development Programme, The United Nations Environment Programme, and the World Bank. Oxford: Oxford University Press, April, 2002.

World Energy Council (1997) Statement to the Third Conference of the Parties to the UN Climate Convention. Press release, Kyoto, Japan.

World Energy Outlook (2005) International Energy Agency. IEA, 2005. Paris, Cedex 16, France. October.

Part VI

CRITIQUE ON TROPICAL ARCHITECTURE

14 IS SUSTAINABILITY SUSTAINABLE? INTERROGATING THE TROPICAL PARADIGM IN ASIAN ARCHITECTURE

Anoma Pieris

Department of Architecture, University of Melbourne

Abstract

Environmental movements, with their arguments for sustainability, provide a cogent critique of late capitalist production. During the 1980s the climatically defined and site-specific architecture of the tropical regions of Asia seemed to have taken on this challenge by presenting a resistance to both the pragmatic solutions of the developmental state and corporate International Style architecture. Architects in South and Southeast Asia actively participated in both the material and discursive dissemination of place-specific forms of regionalism using the "tropical" and "vernacular" categories as related design paradigms. However, faced with the imperatives of providing sustainable architectural environments for their growing urban populations, these very categories began to unravel, exposing the socio-political context out of which they had emerged.

By interrogating the tropical paradigm in architecture, this chapter hopes to identify the challenges to sustainability posed by the complex social relationships that have evolved during the process of decolonizing specific geographies. Viewed within this historic context, the categories of "tropical/vernacular" carry utopian messages of geographic belonging and social equality to a once colonized population. Underscoring this rhetoric is the moral obligation to provide a sustainable physical environment. Under these conditions: is sustainability sustainable? The sites chosen for this discussion are Singapore, Malaysia and Sri Lanka, which have contributed significantly to the above debates by producing a distinctive and highly disseminated "climatic", "regionalist" architecture.

Keywords

Vernacular, tropical, regional, sustainability, architecture.

14.1 INTRODUCTION

Architectural discussions on sustainability are notorious for representing themselves within the confines of empirical studies, scientific rationalism or quantitative analyses that have shaped building technology methodologies over the past fifty years. For these very reasons they often fail to have the necessary social and cultural impact that might enable them to enter the everyday sensibilities of a local culture or inform ordinary social practice. The ideological, imaginative and political significance of environmental sustainability is often ignored, understated or romanticized leading to its marginalization as a specialized area of study. When entrenched in a landscape of positivist analyses an encounter with a sociologist Manuel Castells' assessment of the socio-political terrain of environmental movements is indeed refreshing.

Castells identifies environmentalism as a major social movement in our time and uses a very provocative title "The Power of Identity" to engage debates on environmental sustainability with the identity politics of what he describes as the information age. In his interpretation, environmentalism foregrounds ideas of social justice and creates a space from which to critique the relentless progress of the late capitalist global order (Castells, 1997). In a chapter titled "The Greening of the Self: The Environmental Movement," Castells charts a map of what he describes as the kaleidoscope of environmentalism including nature lovers, local communities, the green self, internationalist eco-warriors and concerned citizens. Each of these groups has specific adversaries: uncontrolled development, polluters, industrialism, technocracy and patriarchalism, unfettered global development and the political establishment, which they encounter on the path to particular goals, namely and in their given order, wilderness, quality of life/health, ecotopia, sustainability and counter power. Sustainability for which the suggested example is Green Peace is identified as the goal of internationalist eco-warriors who militate against unfettered global development.

The objective of this exercise of untangling the distinct strains of the environmentalist discourse is to understand their significance as vigilante groups in a century where the most significant political opposition to capitalist industry in the form of a socialist welfare state is being dismantled. The concept of the "green self" foregrounds the individuation of this approach into a self-conscious system of identity-based, beliefs or practices among the citizens of (largely) first world nations. It is a shift towards a self-reflexive examination of the sensibilities that were severed through industrialization. Conscious of the role architecture has played in producing both the problem and routes of escape from it, we may ask, how could we contribute to this discourse? For the purpose of this essay our question must be directed at Asia, the region that is industrializing and urbanizing its built environment at a scale unprecedented so far.

Throughout their post-colonial history, nations in South and Southeast Asia embraced the developmental ideologies of both the capitalist and socialist models with scant regard for their ecological consequences. With improved levels of health care, post-independence, these societies witnessed a population explosion but neglected to educate successive generations into such a sensibility. While it is possible to trace ecological sensitivities in a number of Asian cultures, for example in pre-colonial practices in rural environments, the social processes of colonization and urbanization have long diminished the influence of these lessons from the past. If ecological sensitivities are expressed by Asian populations they are more likely to be the by-products of poverty, the social economy of frugality rather than a self-conscious recognition of the need to sustain the surrounding environment. As Asia aspires to greater degrees of industrialization and affluence, reliance on an environmental morality shaped by economic need, will certainly fail to generate the desired sensitivity to issues of sustainability.

Therefore our discussion of sustainability must attempt to understand how it might be inserted into a largely hostile Asian urban context and the terms by which it might justify its presence. In short who should be the audience for a discourse on sustainability and why so? In order to formulate this argument this essay will attempt to plot and analyse moments in contemporary history when issues of sustainability were invoked by self-reflexive architectural practice. A few practitioners and debates from Singapore, Malaysia and Sri Lanka have been selected for analysis since, in my interpretation these architects have benefited from such invocation.

14.2 THE TROPICAL CITY AND THE GREEN SKYSCRAPER

Two concepts of the "tropical city" and "green skyscraper" are amongst the most familiar outcomes of environmental consciousness by Asian architects. Their proponents Tay Kheng Soon and Ken Yeang have been vociferous in their defence of what is to a large extent an individual vision (Tay, 1988; akitektenggara.com/ideas/tropical_city_concept, 2005). Tay, who has published extensively on the design of tropical cities, adopts an enlightened yet technocratic approach. He believes that through policy, legislation and co-ordinated urban design many of the conceptual biases of the colonial planning structure might be re-conceptualized, to better suit the requirements of what he describes as "The Tropical Asian City of the 21st century" His reaction is against the outcomes of half a century of a functionalist paradigm where a nation bent on expedient modernization stripped its urban geography of its messy tropical habitat reintroducing it selectively and in a sanitized and ordered fashion in prescribed landscaping. It produced what Cherian George described as "the Air-conditioned nation", "designed first and foremost for the comfort of its inhabitants (George, 2000)." Tay reacts to this prerogative in his manifesto and in the approach of his design practice *Arkitek Tenggara*. In his vision for the twenty-first century Asian city he declares "Clean and sterile buildings and spaces separated from nature and from human activity should be avoided or ameliorated". He calls for an intelligent city, where "the ambient information content has to be increased". Such appeals for environmental sustainability through instrumental solutions and a self-reflexive citizenship suggest and echo the strains of socially directed Modernist agendas from the 1960s, which will be visited later on in this argument. An important part of Tay's approach is in the insertion of ambient microclimates into the dense high-rise environments of Asian cities. His design for the Duxton Plain Public Housing Competition for a 50-storey residential complex, (the very first of that height) in Singapore in 2001 was described as a vertical parkland (akitektenggara.com/ideas/duxton, 2005).

The approach to green verticality has also been championed by Ken Yeang, a Malaysian architect. Over the years he has progressively inserted environmental solutions into his designs for high-rise buildings in a growing repertoire of technologies ranging from concrete to steel and glass. Yeang's architecture, which experiments with passive low-energy solutions for multi-storey buildings, aims to consider "their impacts on the site's ecology, and the buildings' use of energy and materials over it's life-cycle" (trhamzahyeang.com, 2005). Yeang has fashioned his architectural persona via the green dialogue through repeated publications, presentations and experiments with the design of "bio climatic skyscrapers" (Ken Yeang, 2000). Such an approach is particularly unusual for Malaysia where Yeang's practice *Hamzah and Yeang* are based. Skyscrapers are relatively few (Cesar Pelli's Petronas Tower being the most prominent example) and Malaysian cities lack the critical urban density that might *sustain* skyscraper design. Yeang's company profile lists several international projects of around 50 storeys, as skyscrapers.[1]

Studied in this light both these practitioners need to be described as visionaries, eco-warriors imagining urban futures, at a scale which they may not live to see actualized in their own environments. Buildings of these scales have been realized in the neighbouring East Asia. Their foundation is clearly modernist while their association with the region is drawn largely from their iden-tification with and responses to a specific tropical climate. Tay in particular, due to his activism through SPUR (Singapore, Planning and Urbanism Research Group) could be seen as a reactionary voice that broke away from the developmental ideologies of an early nationalist government. However, both these architects subscribe to a capitalist economy, which produces high-rise buildings as the ultimate symbol of their international status. Abidin Kusno's argues that Tay and Yeang have attempted to re-negotiate the authority of the international style by masking the modernist box in an environ-mental filter (Kusno, 2000). He further observes that a very specific form of plural society and political culture frames their efforts at con-structing regional identity via a locally meaningful trans-local climatic discourse. It rejects the primordialism of cultural identities based on visual iconic identification in order to respond to the interna-tional expectations of a first-ranking Asian economy (Kusno, 2000). But the discourse on regionalism and its meaning for these two practitioners is far more complex than this interpretation might sug-gest. It is an outcome of a much longer history of architectural engagements with both climatic and regional identity. Tay, who has discussed the issue, in his monograph, identifies three reasons for the focus on climate over other factors. They are: (a) the pressures from the Malaysian government to produce a Malaysian identity in design (and the need to define one that is not overtly Islamic); (b) a disquiet amongst Malaysian architects at an implied sectarianism in the choice of ethnic symbols; and (c) the tendency to follow Western post-modernism, which he sees as historically absurd and dangerous because it exacerbates ethnic cleavages (Tay, 1989).

14.3   TROPICAL MODERNISM

If we are to trace the lineage of discussions of sustainability and the call for an instrumentalist approach to design for tropical climates by Yeang and Tay we find its roots in a Modernist discourse on "design for tropical environments" that originated in Europe. A few notes on the history of this tradition need to be introduced at this point. As discussed by Lai Chee Kien, a series of conferences held in Europe during the 1950s transformed a colonial and in many ways *orientalist* construct into a rationalized approach and created oppor-tunities by which former colonial powers could continue to intervene in former colonies (Lai, 2002). Lai notes two tropical schools that emerged during that period. The Architectural Association, London commenced it diploma in tropical architecture under

Maxwell Fry in 1954 and from 1957 onwards under Otto Koenigsberger. Seven years later in 1962, the University of Melbourne began a postgraduate course in tropical architecture (Lai, 2002). The systems explored by these programs responded directly to the tropical climate through the functionalist paradigm of a Modernism inherited through a western education. Passive solar design and ventilation systems that were evident in the local vernacular were translated into new materials and technologies such as concrete, steel, glass, asbestos, etc. More importantly it trained the first generation of Asian architects. We see this approach in the work of the *Malayan Architects Co-Partnership*: Lim Chong Keat, Chen Voon Fee and William Lim. As discussed by Philip Bay they explored the tropical paradigm through modern programs: a conference hall, a corporate high-rise, and mixed residential and retail developments (Tzonis et al. 2001). Both Tay Kheng Soon and Ken Yeang (in Malaysia) continued this form of experimentation using ecological agendas. In Sri Lanka the approach is visible in the early work of Minette de Silva and Geoffrey Bawa. The Sri Lankan architect who carried this tradition even further was Valentine Gunasekara (Figure 14.1). We could describe this as the first self-conscious discourse on environmentally sustainable architecture to take root in Asia.

The tropical modernist approach contrasted sharply with existing architectural trends largely formulated by the colonial public works

14.1.
Valentine Gunasekara, Jesuit Chapel, Colombo 1960 (*Source:* Gunasekara, published with permission from the source).

departments. As in the case of Edwin Lutyen's design for New Delhi the P.W.D. produced a particular nineteenth-century revival of syncretic Indic styles wrapped around gothic or neo-classical plans. In the case of the British it was an attempt to translate the monolithic administrative structure imposed upon its colonies through recognizable forms and motifs more palatable to a population who were agitating for independence. The inevitable ethnic incisions caused by the production of a national geography were inscribed in the prescribed architectural styles: Indo-saracenic for Malaysia and Buddhist for Sri Lanka.

The rejection of Indic styles by the tropical modernist architects during the 1960s and 1970s can be interpreted as a move towards secularism, in an environment troubled by ethnic struggles over national identity and territory. At first their architecture appeared to be aligned with the pragmatic functionalism of the developmental state, intent on large-scale projects of modernization. The involvement of many tropical modernist architects in Le Corbusier's design for Chandigarh, serves as an example of this early alliance. Subsequently, their dissatisfaction with state hegemony and rising ethno-national consciousness would erode such alliances.

By distancing themselves from the identity politics of the colonial state and from the socialist pragmatism of the newly independent nation, architects of the tropical modernist school refused to subscribe to the anxious, contentious and often narcissistic ethnic nationalisms that were seeking definition. The approach had many lacunas: its cultural expressions were borrowed from clearly Western: Modernist and Brutalist styles and its programs followed utopian experiments of the period. The links to the displaced metropolitan centres of the tropical discourse in the UK and Australia, while creating opportunities for experts like Le Corbusier, Maxwell Fry, Jane Drew, and many others to design in former colonies, suggested a prolonging of the colonial relationship. Moreover, the tropical modern approach did not account for the environmental costs of manufacturing, transporting and constructing the buildings concerned.

14.4   CONTEMPORARY VERNACULAR

The diminishing influence of the tropical modernist idiom during the 1980s can be linked to three factors: the disenchantment with modernism due to the rise of postmodernism in western architectural discourse, the rejection of modernism as an expression of socialist pragmatism (and communist ideology), and the reaction against utopian projects as vehicles for social engineering associated with the post-war nation-state. Underwriting this failure was a shift in Western audiences away from universalized commodity production towards pre-industrial forms and manufactures. For the Westerner the dirty realist cities and fragile vernacular habitats of Asia preserved the processes that had been sanitized in the First World.

In short, the search for authenticity in the post-industrial nations of the West would create the opportunities for a vernacular revival in Asia providing it with its keenest audience. Ironically it was the self-same image of authenticity that had been exploited by nineteenth-century *orientalists*. In a century of colonial exhibitions, it had been used to mark Europe's comparative progress against a version of Asia, which was repeatedly reconstructed as a world of thatched huts and friendly savages. As Asia began to modernize, the vernacular was similarly charged with numerous associations, in this instance where pre-colonial/primitive cultures were being appropriated by bourgeois consumers.

Although the contemporary Asian vernacular appears congruent with the post-industrial architecture of post-modernism in terms of its reference to cultural, ethnic, historic and mythical narratives, it differs sharply in one respect. Its products reveal labour-intensive processes, where the mark of the individual artisan is inscribed onto the material surface. The rusticated appearance of vernacular architecture is grafted onto a modernist plan into which labour saving appliances and sanitized facilities have been inserted. Its consumers, wealthy elites and new bourgeois groups with strong urban sensibilities construct the tropical vernacular as a refuge from the dusty, contemporary Asian city outside. During the 1980s, the climatic relationship was being used to construct a vision of a rural utopia that was being rapidly abandoned by Asia's poorer citizens.

My objective in analysing the contemporary vernacular discourse is to understand how its invocation of climate, local geography, local materials and construction methods generated a mirage of an environmentally sustainable practice. How the rising sensibility of the "green self" in the West, contributed to a form of design morality that was referenced through the tropical vernacular. Its strongest defence came from Sri Lanka where the publishing industry and western discourse collaborated to produce the work of a single architect, Geoffrey Bawa, as an iconic figure of a vernacular revival (Figure 14.2).

The adoption of the vernacular as a source for contemporary design in Sri Lanka has a history that predates the work of Geoffrey Bawa and is found in the work of Sri Lanka's first woman

14.2.
Bawa: DeSarem Houses,
Colombo (*Source:* Author).

architect Minette de Silva. She, like Bawa and Gunasekara was trained in the AA school (de Silva, 1998). Minette's primary interest was in how a society and culture used and understood space, and her architecture was adapted to suit this criteria. Her sources were the vernacular buildings of the rural landscape of Ceylon. Bawa converted to this orientation in the late 1960s and began to introduce rural courtyard architecture into Colombo, the colonial capital, a landscape dominated by formal colonial bungalows. His approach clearly differentiated itself from the architecture of the nation-state, which, following the example of the colonial public works department, produced a hybrid Buddhist/neo-classicism (Figure 14.3). It alienated Sri Lanka's ethnic minorities, during the critical years following independence.

Bawa's rejection of modernism also responded to the pressures of a larger political and economic climate. Sri Lanka, like India, had chosen to follow a markedly socialist economic direction and embarked on a policy of import substitution. Among other things this led to a dearth of cement products which was exacerbated when, following the OPEC oil crisis of 1973, several other imported construction materials went out of circulation. The OPEC oil crisis and the boom of the oil economies that followed it produced a particular architectural counter culture championed by the Aga Khan Program. It was created in reaction to the epidemic of international style architectures that overtook nations in the Middle-East. Through its *MIMAR* journals,[2] annual awards and design monographs, the Aga Khan Program formulated its manifesto for sustainability directed at local communities living in vernacular

14.3.
Independence Hall, Colombo
(*Source:* Author).

environments. Its defence of local construction technologies and social and cultural practices of the less affluent communities in Islamic countries won it a large audience in the developing world. The *MIMAR* monograph on Geoffrey Bawa was published in 1986 (Taylor, 1986).

Bawa's revival of the local vernacular attracted many associations. On the one hand, it resembled the cultural artefacts of the pre-colonial past and was hailed as a patriotic gesture. On the other hand it drew from the colonial vernacular, particularly the island's own Portuguese tradition and other, Indian, Italian and Spanish vernaculars, reflecting a truly post-modern approach. It won him many students and allies in the next generation of architects resulting in their rejection of modernist experiments. But Bawa's clientele digressed considerably from the rural communities publicized through *MIMAR*. They were largely from the Colombo elite, who were self-consciously re-orienting away from their colonial habits towards new strains of national expression.

In view of its complexity it is important to evaluate the ethics underlying the contemporary vernacular. In contrast with the socialist modernism or ethno-centric expressions of the nation-state the simplicity of the vernacular tradition, its climatic sensitivity, rural roots and place-based construction practices, seemed the most appropriate repository of everyday Sri Lankan architectural expression. It drew its strength from the moral economy of rural life that valued and protected the interests of Sri Lanka's majority rural inhabitants. Its rhetoric was seen to be accessible to local

audiences. In this interpretation the vernacular was the most democratic expression of the new nation. Its contribution to sustainable
practices was an unself-conscious by-product of these several
attributes, it was ostensibly an architectural common ground where
the life-styles of local rural communities would be defended by concerned bourgeoisie citizens. However, it would not maintain this
particular relationship for long.

The dissemination of the designed vernacular among Colombo's
elites gave it a very different meaning and created attendant and
problematic associations. Colombo's elites were nostalgic for pre-
colonial *walawwas*: sprawling residences comprised of courtyards
and verandahs, where their ancestors had lived as village headmen.
British plantation bungalows had also borrowed from these forms
and developed a leisurely lifestyle modelled on feudal practices
(Figure 14.4).

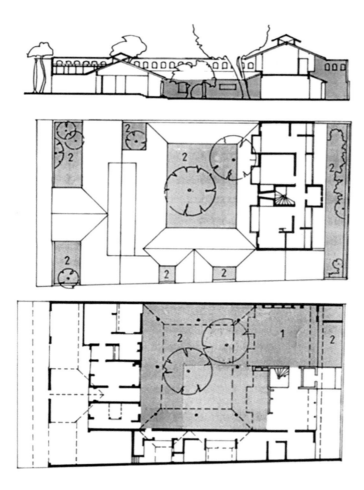

14.4.
Bawa, Ena de Silva House
(*Source:* Author, based on
drawings in Taylor, 1986).

The ideology of the Victorian picturesque when grafted on to the *walawwa* formula produced a particular sensibility: the leisurely lifestyle of the landed gentry with their scenographic appreciation of the natural environment. Tzonis and Lefaivre have identified this approach as "Romantic Regionalism" (Tzonis et al., 2001). Not only did the homes commissioned by Bawa's clients fail to express the moral economy underlying the vernacular impulse, they were produced at considerable cost through labour-intensive processes. They also consumed large quantities of local materials, like clay-tiles and timber. Its environmental consequences are best conjectured in the following reference based on 1978 figures.

"... in 1978, there were some 3000 recognized brick kilns in Sri Lanka alone, each producing an average 150 000 bricks a year, sufficient to build over 50 000 houses. But not without cost to the consumer, and a heavy one to the environment: a firing of 25 000 bricks can consume forty forest trees, a rate of deforestation in Sri Lanka alone of around 750 000 trees a year" (Oliver, 1987)[3].

The most detrimental impact of this approach would be felt when the vernacular was inflated to suit a modern program: a hotel, university, or parliament. In the service of large urban and national projects, such as these, which were commissioned under political patronage, the contemporary vernacular reached beyond its local confines to win an international profile as a kind of regionalism.

14.5 CRITICAL REGIONALISM

In their introduction to *Tropical Architecture: Critical Regionalism in the Age of Globalization*, Alexander Tzonis and Liane Lefaivre outlined the evolution of the regionalist approach in the post-war period (Tzonis et al., 2001). They described how initially, regionalism was identified with nationalist totalitarian tendencies and rejected by proponents of the International Style who, after a decade of war, sought architectural unification through universal images. It was Lewis Mumford's conceptualization of regionalism as a humanist resistance to mechanistic functionalism that reinstated local subjectivity as a starting point for architectural exploration (Mumford, 1967–1970). This subjectivity could be expanded to explore sustainability at many levels. The authors' arguments drew on concurrent philosophies: the critical theory of the Frankfurt School, Victor Shklovsky's ideas of de-familiarization and phenomenological explorations of place-based experience (Tzonis et al., 2001). It was out of this critical stance that an architecture evolving from locale was reframed as ecological and community based and took on its emancipatory rhetoric. Critical regionalism was initially framed by Kenneth Frampton as a resistance to the universalizing international style architecture that continued to service post-industrial urban environments.

Throughout the post-colonial period, South and Southeast Asia emerged as the most intense site where regionalism was produced discursively, textually and architecturally. Its circulation was originally validated through its vernacular derivative and consequent accessibility as a common imaginary for the majority of the population. As such its characteristics: response to tropical climate; social habits such as outdoor living; material and tectonic transparency; and embeddedness in rural origins; constructed a system of values by which to judge post-independent attitudes and was adapted to suit modern or progressive aspirations. As described by Philip Bay in his essay titled "Three Tropical Design Paradigms" it had many manifestations that could be seen as progressive, romantic, or iconic kitsch (Tzonis et al., 2001).

At its best, regionalism gave confidence to local identities by circulating images that were familiar to rural populations in South and Southeast Asia but at its worst these images represented the same positions of privilege that had once been occupied by colonial residents. In land-scarce Singapore, rural Malaysia or poverty stricken Sri Lanka, where the majority of citizens lived in very different forms of housing (mud huts, slums, tenements, kampongs, Housing Development Board Apartments) such architectural examples ultimately captured a utopian dwelling, which was unavailable to the rest of the population. Both the vernacular and the tropical had been re-territorialized by the westernized elite as an exclusive and picturesque terrain.

Architect Geoffrey Bawa, along with many others became passive subject of this discourse during the late 1980s when, following projects in Batujimbar Bali, Bawa's residential architecture was disseminated throughout the Southeast Asian region. It captured and marketed the colonial/feudal ambience, in its vernacular trappings, to an international audience. Since resort hotels rely on picturesque settings as their major attraction it proved an ideal program for this particular architectural approach. Not only did the resorts depend on *orientalist* constructions of local culture as exotic, subservient or sensual, in order to bait western tourists but also they were patronized by the local elites who subscribed to these constructions. The resort hotel was inserted into the rural landscape as a gated community that celebrated a deliberately designed vernacular. The local villagers entered this environment as labour, playing the part of colonial servants to tourists and urban elites. Far from decolonizing class, the new vernacular had reproduced colonial/feudal social relationships and emphasized class divisions.

Multiplied at the rate of several hundred hotel rooms per project the use of local materials to produce a vernacular ambience would also have serious ecological consequences.

The tensions underlying the transformation from vernacular-derived architecture to regional icon is best expressed in Geoffrey Bawa's parliament building in Sri Jayawardenepura. The building complex stands on an island surrounded by an artificial lake dredged out of the wetlands adjacent to Colombo. It was the first symbolic move in an effort at reclaiming Colombo's inundation plains for urban development. The building is arranged as a series of pavilions, derived from the democratic structures of wayside resting places but is separated from its public by its artificial moat. The vernacular language of the pavilion form and the local technologies it implies has been abandoned in the quest to produce a much larger building complex that serves a modern program.

Whereas the objective of critical regionalism was to identify specific practitioners as concerned citizens of a particular locality, the proponents and practitioners of regionalism had reduced their consumers to arm-chair nature lovers. The poverty outside the hotel walls would be carefully concealed or would merge into a picturesque backdrop, which the Western tourist would not have to encounter. The parliament was typically depicted through aerial photographs, which would maintain its magnanimity from the desired distance. The sanitization of the vernacular experience and its flattening into a two-dimensional backdrop ultimately encouraged a particular scenographic appreciation of the tropical climate. Its audience were no longer the local community but a community of global consumers. Their consumption of the picturesque became most evident in Singapore during the 1990s when it produced an uncritical tropical architecture.

14.6   NEO-TROPICAL ARCHITECTURE

Whereas Sri Lanka's elites had no intention of abandoning their class privileges, post-independence Singapore's social make up was altered by economic democratization. Here, the socialist model was married to a progressive economic policy intent on modernization. Singapore needed to effect the harmonious distribution of three distinct ethnic categories (Chinese, Malay and Indian) in an urban landscape and it chose to do so through residential redistribution. The means of its execution, which involved housing its entire population in high-rise apartments, physically removed the citizens from their natural geography and increased their reliance on man-made structures and services. Indeed Singapore gained

14.5.
Kampong House, Malaysia.
(*Source:* Author)

considerable notoriety for the speed with which it was flattened, paved over and appropriated land for development (Figure 14.5). The consequence of this accelerated progress was that the Singapore vernacular in the form of a village (kampong) and the so-called Malay house (a timber house on stilts) became a thing of its past, found only in neighbouring Malaysia or Indonesia and was regarded as a sign of underdevelopment. Kampong life, kampong habits and kampong architecture was alluded to as a metaphor for a previous chaos, which had been replaced by the progressive modernity desired by all Singaporeans. Even as the kampong image was pushed across the border into Malaysia it was reproduced as a nostalgic subject, a lost hinterland wherein vernacular origins were rooted.

Malaysia reproduced the tropical vernacular through a similar nostalgia, for a kampong community located in the rural outskirts of the city. With modern super-highways that compressed the physical distance between city and its hinterland, Malaysia's bourgeoisie could relocate themselves outside its dusty periphery in redeveloped palm-oil plantations. These modern kampongs were designed as gated communities secured by high walls and security posts from the poorer kampongs adjacent to them. Ordinary Malaysian citizens could choose between existing rural kampongs, urban slums and kilometre upon kilometre of identical urban terrace houses.

During the 1990s, Singapore emerged the hub of a publishing industry that marketed the tropical image, the Southeast Asian region and the modern Asian life-style to its global consumers (Powell, 1993, 1996, 1998, 2001; Tan, 1994, 1996, 2000; Marsden, 2002). Not only was such a life-style exclusive in this context it envisioned a Utopia denied to all but its affluent citizens. The new "tropicality" that was reshaped in modern Singapore had its own attendant architecture. Throughout the tropical belt of the Southeast Asian region the vernacular had become the pre-occupation of a wealthy bourgeoisie who would weave references to the lost vernacular into sanitized tropical homes, designed after colonial bungalows, at great cost. In them the act of living on landed properties, of owning a piece of the national geography rather than a faceless urban terrace house or a floating slice of real estate signalled the enormity of each nation's sacrifice to modernization. On reading the annual list of architectural publications out of Singapore, with their ubiquitous neo-tropical bungalows, one would never guess the extent to which the actual context and its architecture were being under-represented.

If we are to reconsider the neo-tropical approach in terms of Castell's interpretation of the "green self" we find that it was largely predicated on maintaining the critical separation between the lived experience and the scenographic backdrop that might make tropical living tolerable. Issues of sustainability had been abandoned in the process of commodifying the user's relationship with the natural environment. Sequestered within the air-conditioned environment the post-modern gaze to the tropical wilderness, being cultivated in the garden beyond it, was one of uncritical consumption. In this final image the fundamental premise of Castell's argument had been co-opted into the vast machinery of late capitalist commodity production. By the end of the twentieth century the tropical paradise would become the most marketable image via which Asia could insert itself into the global marketplace. It would do so at any cost and by any means available.

A strident voice against the reworking of the tropical debate during the 1990s was that of Tay Kheng Soon, who was dismayed at its descent into a fashion statement. He claimed that neo-tropicality (a term that was coined for this approach), was a form of neo-colonialism slavishly derivative of the cubic and planar geometries of the modernist avant-guarde. In his view, recycling modernism via a forty-year-old debate, did not constitute a new approach, "And so my regret is that the new style defers and deflects the quest. The quest for a contemporary architectural aesthetic of tropicality in our own terms and none other" (Tay, 2001).

14.7 CONCLUSION

The above statement by Tay forces us not only to revisit Kusno's critique but also to rethink the definition of sustainability outlined so far. On the one hand, sustainability must be expanded to incorporate lessons in citizenship and in critical resistance to a prescribed aesthetics derived from the West. In order to succeed it must be incorporated and written into the existing political economy and must not be allowed to replicate itself indiscriminately as an object of desire. Missing from this argument and essential to Castells' analysis of environmental movements is the critique of the capitalist machinery that endlessly reproduces these several tensions.

As implied in my introduction, the folding-in on itself of the discourse on sustainability imagined through climate and the difficulties faced by a new breed of eco-warriors stem from a weak link in the quest for a more sustainable environment. Critiques of global capitalism cannot be successfully launched by a profession that services the capitalist machine, and is in fact its main champion. In order to transform our built environment in the pursuit of environmentalist goals in Asia, we would have to convince the construction industry to abandon several profit-making practices. Educating our consumers in self-reflexive practices is a task that our profession has not undertaken so far. As we participate in the ever-increasing urbanization of Asia the most that we can do is to modify the end product so as to limits its detrimental impact on our ecology. Our interventions so far have been relatively benign. If a counter-culture that truly speaks the language of environmental sustainability is to emerge from architectural practice it will be one that does not build but simply intervenes, modifies and subverts existing building practices.

NOTES

1 What indeed is a skyscraper? Although during the early twentieth century it was used to describe several high-rise and multi-storey buildings these would no longer satisfy a contemporary definition of the type. For example the Woolworth Building by Cass Gilberts, was the tallest building in New York at 57 Storeys from 1913–1930 when this position was usurped by William Van Alen's Chrysler Building at 77 floors.
2 *MIMAR* journals were published from 1981–1992.
3 In reference to David Robson, Gormley and Sonawane 1984, *Aided self-help housing in Sri Lanka 1977–1982,* report for Overseas Development Administration, London.

REFERENCES

Bay, J.H. (2001) Three Tropical Design Paradigms, in Tzonis et al. (eds.) *Tropical architecture: Critical regionalism in the age of globalization*. Chichester: Wiley-Academy, with Fonds, Prince Claus Fund for Culture and Development, The Netherlands, pp. 229–265.

Castells, M. (1997) *The Information Age: Economy, Society and Culture: Power of Identity Vol 2* (The Information Age: Economy, Society & Culture), Malden, MA: Blackwell.

de Silva, M. (1998) *The Life and Work of an Asian Woman Architect*, Colombo: Smart Media Productions.

George, C. (2000), *Singapore: The Air-conditioned Nation*, Singapore: Landmark Books.

Kusno, A. (2000) *Behind the Postcolonial: Architecture, Urban Space, and Political Cultures in Indonesia*, London: Routledge.

Lai, C.K. Tropical Tropes: The architectural politics of building in hot and humid climates, presented at '*(Un)bounding Tradition: The tensions of borders and regions*', 8th IASTE International Conference, Hong Kong, 12–15 December, 2002 (unpublished paper).

Marsden, J.D. (2002) *New Asian Style: Contemporary Tropical Living in Singapore*, Singapore: Periplus Editions.

Mumford, L. (1967–1970) *The Myth of the Machine*, New York: Harcourt Brace.

Oliver, P. (1987) *Dwellings: The House Across the World*, Oxford, Phaidon.

Powell, R. (1993) *The Asian House: Contemporary Houses of Southeast Asia*, Singapore: Select Books.

Powell, R. (1996) *The Tropical Asian House*, Singapore: Select Books.

Powell, R. (1998) *The Urban Asian House:Living in Tropical Cities*, Singapore: Select Books.

Powell, R. (2001) *The New Asian House,* Singapore: Select Books.

Tan, H.B. (1994) *Tropical Architecture and Interiors: Tradition-based Design of Indonesia, Malaysia, Singapore, Thailand,* Singapore: Page One Publishers.

Tan, H.B. (1996) *Tropical Retreats: The Poetics of Place*, Singapore: Page One Publishers

Tan, H.B. (2000) *Tropical Paradise*, New York: HBI.

Tay, K.S. (1988) The Tropical Asian City for the 21st Century; 'The Intelligent Tropical City as a Framework for Architecture and Planning' in a seminar, Heritage and Change in South-East Asian Cities, South-East Asian Study Group and the Aga Khan Program at Harvard University and MIT, Singapore, Jan., 1988.

Tay, K.S. (1989) *Mega Cities in the Tropics: Towards an Architectural Agenda for the Future*, Singapore: Institute of Southeast Asian Studies.

Tay, K.S. (2001) *Neo Tropicality* Singapore. http://www.akitektenggara.com/articles/2001/neo.htm, [28/03/05].

Taylor, B.B. (1986) *Geoffrey Bawa*, A MIMAR book, Concept Media, Singapore.

Tzonis, A. Lefaivre, L. and Stagno, B. (eds.) (2001) *Tropical Architecture: Critical Regionalism in the Age of Globalization.* Chichester: Wiley-Academy, with Fonds, Prince Claus Fund for Culture and Development, The Netherlands.

Yeang, K. (2000) *The Green Skyscraper: The Basis for Designing Sustainable Intensive Buildings*, Munich: Prestel.

http://www.akitektenggara.com/ideas/tropical_city_concept/tropicalcity.htm, [04/07/2005] Tay, Kheng Soon, The Tropical Asian City for the 21st Century; 'The Intelligent Tropical City as a Framework for Architecture and Planning', Jan,1988.

http://www.akitektenggara.com/ideas/duxton/duxtonpanel.htm, [28/03/05]. The Duxton Plain Project proposal.

http://www.trhamzahyeang.com/profile/company.html, [04/07/2005]. Hamzah and Yeang profile.

INDEX